W9-CXZ-938

Blood Substitutes

Physiological Basis of Efficacy

Blood Substitutes

Physiological Basis of Efficacy

R.M. Winslow
K.D. Vandegriff
M. Intaglietta
Editors

Birkhäuser   1995
Boston • Basel • Berlin

Robert M. Winslow
School of Medicine
University of California
San Diego, CA 92161

Marcos Intaglietta
Department of Applied Mechanics
and Engineering Sciences
University of California
San Diego, CA 92161

Kim D. Vandergriff
School of Medicine
University of California
San Diego, CA 92161

Library of Congress Cataloging-in-Publication Data
Blood Substitutes: Physiological Basis of Efficacy / Robert M.
   Winslow, Kim D. Vandergriff, Marcos Intaglietta, editors.
         p.   cm.
   Includes bibliographical references and index.
   ISBN 0-8176-3804-0 (hard : alk. paper). -- ISBN 3-7643-3804-0
   (hard : alk paper)
      1. Blood substitutes--Congresses. I. Winslow, Robert M., 1941-
. II. Vandergriff, Kim D.   III. Intaglietta, Marcos.
      [DNLM : 1. Blood substitutes.   2. Hemoglobins--metabolism-
   -congresses.   3. Oxygen--metabolism--congresses. WH 450 C976 1995]
   RM171.7.C87   1995
   615'39--dc20   95-2118
   DNLM/DLC   CIP

Printed on acid-free paper

© 1995 Birkhäuser Boston   *Birkhäuser* ®

ISBN 0-8176-3804-0
ISBN 3-7643-3804-0

Typeset by the editors
Printed and bound by Braun-Brumfield, Ann Arbor, MI.
Printed in the U.S.A.
9 8 7 6 5 4 3 2 1

Table of Contents

Chapter 10

Chapter 11

Chapter 12

Chapter 13

Foreword

This volume contains a collection of essays by selected authors who are active in the field of blood substitutes research or closely allied disciplines. These essays were delivered as lectures by the authors at the second annual "Current Issues in Blood Substitute Research and Development - 1995" course sponsored jointly by the Departments of Medicine and Bioengineering, University of California, San Diego, the National Institutes of Health (NHLBI), and the U.S. Army on March 30, 31, and April 1, 1995 in San Diego.

This course had three goals: to present fundamental discussions of scientific issues critical to further development of artificial oxygen carriers, to provide academicians a forum to discuss their current research, and to provide the companies involved in developing products the opportunity to update the audience on their progress. The organization owes much to the solicited comments of the attendees of the 1994 course.

We would like especially to thank the U.S. Army, particularly through the efforts of COL John Hess, who provided significant funding to make publication of this volume possible. In addition, a number of the participating companies provided additional financial support to offset the costs of the course. These include Alliance Pharmaceutical Corp., Hemosol, Nippon Oil and Fat, Northfield Laboratories, and Ortho Biotech.

We owe an enormous debt to Renée Schad, who handled almost all of the formatting and final editing of the manuscripts. As usual, Shirley Kolkey and Complete Conference Management have provided competent, professional guidance and organization for the course. We would also like to thank each of the contributors who worked cheerfully (more-or-less) with us and our rigid deadlines.

Robert M. Winslow, M.D.
Kim D. Vandegriff, Ph.D.
Marcos Intaglietta, Ph.D.

Preface

Although a substitute for human red cells has been sought for more than a century, still no product is available to patients. Until the early 1980's, research and development in this area was relegated to rather obscure academic efforts, but AIDS suddenly brought new focus to the effort when its transmission by blood transfusion was shown clearly. Although blood is now extensively tested for the AIDS (HIV) virus, research on red cell substitutes has shown tremendous potential application for these products, and development continues at an unprecedented pace. In fact, a number of products are currently in human clinical trials.

1994 was an very exciting year in the research and development of red cell substitutes. The year began with a unique conference held in Bethesda, Maryland, jointly sponsored by the FDA, the NIH, and the DOD. An unprecedented number of international conferences were held subsequently, attesting to the growing scientific and commercial interest in this field.

The FDA/NIH/DOD conference set the tone for the discussions which followed: while work and interest continued in understanding of safety and toxicity of existing and new products, more attention was focused on issues of efficacy and on the design of clinical trials.

The content of the course "Current Issues in Blood Substitute Research and Development - 1995" reflects our own research interests and the direction of our UCSD program. Thus, we have selected authors and speakers who can contribute the main theme of this year's course: mechanisms of O_2 transport by red cell substitutes and the implications of these mechanisms for efficacy and toxicity of products.

The volume begins with a fresh look at the world-wide impact that a red cell substitute product would have, and a consideration of allied technologies like blood sterilization and the development of "current good manufacturing practice" (cGMP) standards on the blood banking industry. Economic considerations will determine, to some extent, the commercial success of licensed products, but also in Dr. Tomasulo's opinion, new products will have a tremendous impact on emerging countries whose blood banks are not well developed.

Closely related chapters by Fratantoni (2) and Winslow (3) are aimed at issues of efficacy demonstration. Dr. Fratantoni distinguishes between "efficacy" (the capacity for a product to do something useful for a patient) and "activity" (the capacity for a product to do something physiologically, like carry O_2). Dr. Winslow's contribution describes the broad physiological basis for the "transfusion trigger" — those events which lead a physician to transfuse red blood cells. The chapter makes the case that these triggers are not always clearly defined, therefore definition of clear-cut clinical end-points for red cell substitutes will not be easy. In the chapter by Bowersox and Hess (4) development of red cell substitutes is seen from the unique perspective of the military.

Chapters by Kaufman (5), Manning (6), and Rudolph (7) review the current state of perfluorocarbon-based, hemoglobin-based, and encapsulated hemoglobin products. These authoritative chapters provide readers with a fundamental understanding of the current products undergoing clinical testing and point to future developments and refinements in the products.

Specific issues of toxicity of cell-free hemoglobin are discussed in the contributions by Vandegriff (8) and Blantz (9). Dr. Vandegriff discusses the impact of chemical and genetic modification on hemoglobin stability in regard to heme-globin linkage, oxidation, and the potential for related *in vivo* toxicity. Dr. Blantz describes his recent research in which detailed studies of the effects of hemoglobin solution on kidney function have been carried out. Dr. Blantz goes beyond the traditional studies of gross kidney function and considers hemoglobin-NO interactions and the possible interactions with adrenergic and hormone control of the kidney.

The chapters by Vandegriff and Winslow (10), Intaglietta, Kerger and Tsai (11) and Johnson and coworkers (12) are the most closely related to our own research program at UCSD. In essence, we believe that increased O_2 availability results from the even distribution of O_2 in the plasma containing cell-free hemoglobin. Combined with lowered viscosity of the solutions, this results in autoregulatory vasoconstriction and reduced capillary perfusion. Vandegriff and Winslow discuss the theoretical basis for these views, Intaglietta, Kerger and Tsai describe direct measurements of O_2 distribution in the capillary circulation of the awake hamster, and Johnson and coworkers place these views and observations in the perspective of several decades of research on the mechanisms of autoregulation. If these views prove to be correct, in our opinion, we must reconsider our basis assumptions about viscosity, P50 and, possibly, encapsulation of hemoglobin within an artificial membrane.

Finally, we are indeed fortunate to have Dr. Suit contribute a chapter (13) to one exciting application of cell-free O_2 carriers: the treatment of cancers by enhancing sensitivity to irradiation by increasing tissue PO_2. Dr. Suit reviews basic biology of tumor cell microcirculation and the rationale for using the new solutions.

We hope the readers of this volume and those who attend "Current Issues in Blood Research and Development - 1995" find these chapters as stimulating as we did in assembling and editing them. The selection of topics reflects our editorial biases about the importance of the problems they address, positions not necessarily endorsed by any of our institutional or commercial sponsors. If we have overlooked any areas that should receive more attention, the fault is entirely ours, and we hope the continued feedback from the participants in the course will help keep us attuned to the mainstream of thought in this ever-changing field.

Robert M. Winslow
Kim D. Vandegriff
Marcos Intaglietta

Chapter 1

Transfusion Alternatives: Impact on Blood Banking Worldwide

Peter Tomasulo, M.D.

TM Consulting Inc., 1440 Laburnum Street, McLean, Virginia 22101

1.1 Introduction

The purpose of this chapter is to present an overview of current blood banking organizations and activities and the effect of the introduction of red cell substitutes. To do this job well, one would need to know the characteristics of each licensed red cell substitute, the sequence of licensure and their applications, costs, side effects, *etc.* One would also like to know what other new technologies will have been introduced at the time of the introduction of red cell substitutes. Of course, none of this information is available today and the range of possibilities is very broad. Therefore, any prediction of the impact of transfusion alternatives on blood banking must be considered less than precise. However, it is possible to focus on trends and principles to help the reader make predictions as s/he observes changes in the marketplace and as transfusion alternatives come closer to being a reality. To be inclusive, some circumstances that are not considered likely will be introduced and analyzed. This should prepare the reader a little better if today's opinions turn out to be incorrect.

The approach includes a description of the status of Transfusion Medicine. The best data and the greatest chance for accuracy are in the description of North America because the author is less familiar with the situations in Europe, Japan and the rest of the world. However, some clear trends and circumstances will be presented. The description of the status will be followed by a brief inventory of forces acting on Transfusion Medicine. Politics, technology and the interests of worldwide health service organizations will be considered. The possible impacts of red cell substitutes in different parts of the world will be evaluated.

Blood Substitutes: Physiological Basis of Efficacy
Winslow et al., Editors
© Birkhäuser Boston 1995

1.2 Overview of Status of Blood Banking Worldwide

1.2.1 The United States

Transfusion Medicine has been a relatively conservative and stable part of the U.S. health care system. The basic technologies were introduced 40-50 years ago and have not been changed. What has improved is the testing and sorting of therapeutics for unwanted characteristics. But the therapeutic efficacy of most blood components has not changed and, interestingly, the therapeutic efficacy has not been well-defined for red cells. Transfusion Medicine laboratory activities remain more labor intensive than is desirable, and the technology in the blood bank/crossmatch laboratory is quite primitive in comparison to the rest of a modern clinical laboratory.

The AIDS epidemic has created turmoil for blood bankers, blood bank regulators and for those who are accountable to the public for the quality of the blood supply. It is impossible to summarize precisely what the American people think about their blood supply and there are conflicting impressions. On the one hand, there continues to be remarkable support from blood donors and the companies and organizations with which they are associated. Many recipients and potential recipients seem to understand the very low level of risk and accept it. On the other hand, some segments of the U.S. public are willing to believe that their blood supply is not as safe as it ought to be; not even as safe as the blood bankers say it is. Inquiries into the present and past practices concerning the safety of the U.S. blood supply have been ongoing at the request of Congress and the Department of Health and Human Services. There has been for the first time in modern memory a clear lack of confidence in blood bankers and in the safety of the blood supply. Most people have not lost confidence, but there are enough concerned people that this lack of confidence is a problem of the greatest significance for U.S. blood bankers.

Americans do not like the nature of the risk associated with blood transfusion today. They want more information about blood and they want to gain control over the risk themselves. And when it is possible, they want blood transfusion with no risk at all. It is this desire that has alerted elected officials and U.S. regulatory agencies.

The response of the regulatory agencies has been to increase their efforts to force blood bankers to adopt the procedures used by drug manufacturers. For some time, the U.S. Food and Drug Administration (FDA) has considered blood components to be drugs and it regulates U.S. blood centers in the same fashion that it regulates drug manufacturers. The FDA is implementing a strategy which it believes will raise the U.S. blood centers' level of performance to that achieved by drug manufacturers. Blood center professionals accept the classification as manufacturers and endeavor to meet the standards precisely as the FDA wishes to apply them. There are some differences between drug manufacturing and the provision of blood components, which may make it difficult to achieve the desired status.

Drug manufacturers try to make absolutely pure compounds with a small number of components which have well-understood effects. Compared to most drugs, blood is complicated and nonspecific. It has too many components and too many effects (desirable as well as potentially dangerous). It is known that many substances in blood prepared for transfusion can have no beneficial effect on the recipient and therefore should be removed to prevent them from causing harm. There is currently no technology available which allows blood manufacturers to remove unwanted components to the degree possible in drug synthesis. Blood filtration to remove white cells is a rudimentary step in that direction.

In manufacturing drugs, raw materials must be qualified. This qualification is the most important step in the "manufacture" of blood components. It is difficult to imagine that drug manufacturers would ever depend on interviews of thousands of volunteer individuals over whom they have no control for the quality of their raw materials. Nor would they release hundreds of lots of final product each day. There is no alternative for transfusion medicine.

Drug manufacturers must demonstrate safety and efficacy before their products are licensed, but today there is no agreement on the appropriate efficacy measurements or even the specific indications for some blood components. Drugs can be subjected to lot testing before release, but the FDA defines each blood component as one lot. Because of sterility and other concerns, blood components cannot undergo the same end-product testing.

Drug manufacturers are required to list all components (active and inactive) of licensed agents. Blood centers cannot certify that there are no extraneous chemicals or infectious agents in blood, because their data about the donor (the real manufacturing plant) could never be as complete as the data available to drug companies. Blood centers do not know what the donor ate or to what s/he was exposed. Drug companies have contracts and issue specifications to their raw material suppliers. In contrast, blood centers cannot hold blood donors accountable for complete knowledge about the blood they provide or for maintaining a steady state. Donors who do not know they are in the early stages of AIDS or hepatitis infection provide an example of individuals who might believe they are healthy when they give blood, but who will discover sometime later they are not.

Although U.S. blood centers support the application of the drug current Good Manufacturing Practices (cGMP) and are making every effort to comply therewith, it is clearly more difficult for blood centers to do this successfully than it has been for the pharmaceutical manufacturing companies. Blood centers cannot precisely match the procedures and the achievement of drug manufacturers. There are more variables and there is less potential for control. This added degree of difficulty has and will continue to lead to major issues between the regulated organizations and the regulators. These issues have exacerbated the lack of trust in the blood providers exhibited by some of the American public. In the current environment, blood banking professionals are being forced to concentrate their efforts on regulatory compliance to improve blood safety.

One result of the concentration on regulatory compliance has been that blood banking organizations have not allocated the resources and attention necessary to maintain and increase the quality of customer service. As a result, the adequacy of the homologous blood supply has suffered. Maintaining the participation of donors has become more difficult because necessary cGMP procedures require the elimination of some safe donors, and new screening and education procedures have alienated others. There is currently great difficulty in meeting the needs of hospitals in some U.S. metropolitan areas. Also, because of attention to regulatory compliance, there has been a decrease in the rate of development of new technologies that might improve the adequacy and safety of the blood supply.

1.2.2 The Developed Countries

Much of what has been happening in the U.S. is also happening in Europe and Canada. A segment of the population is very uncomfortable with blood safety and current blood center management. Efforts to eliminate all risk from transfusion have been called for as vigorously in Canada, Japan, France and Germany as in the U.S. There have been official investigations of the safety of the blood supply and recent policies and procedures used by blood banking professionals in Canada, Germany, France and Switzerland. Regulatory agencies and blood centers outside the U.S. also have accepted the benefits of the current drug Good Manufacturing Practices.

Nonetheless, in some countries, blood centers have avoided the inward focus that has afflicted U.S. blood centers. There has been a less dramatic effect on the adequacy of the blood supply and customer service. In fact, some European blood centers seem to have continued innovation so they may be ahead of U.S. blood centers in some respects. France, for example, has set a standard for leukodepletion at $<1 \times 10^6$ remaining white cells/component, which is better than the standard set by many American organizations (Chassaigne 1994). In addition, techniques to prevent alloimmunization have been applied more vigorously in some European countries than in the U.S. The result is that some blood bankers contrast the rate of alloimmunization in their country quite favorably with that seen in the U.S. (Brand 1994). Blood banking professionals and scientific/technical experts in other countries might have more resources and the opportunity for more creativity than exists in some U.S. blood banking organizations.

1.2.3 Developing Countries

Transfusion medicine services are in very short supply or not available at all in many developing countries. A supply of volunteer blood donors is not always available, and when it is, the health of the donors is not always ideal. The infrastructure to recruit and maintain a high quality donor panel is expensive and, in some ways, quite sophisticated. It is not surprising that in developing countries, the blood supply has not attracted the needed resources. The testing, sorting and labeling system which is used so successfully in Europe, Japan and North America, does not lend itself to easy implementation in communities without experienced technologists, a steady flow of funds and with a high rate of positivity in the tests done to select acceptable units of blood. The lack of a Transfusion Medicine system must lead to serious deficiencies throughout the rest of the health care delivery system. It is likely that there will be an increase in demand for safe blood when health care develops in these countries. If surgery is to be performed and malignant disease treated, these countries will need a way to replace missing blood components.

1.2.3.1 Global Implementation of Good Manufacturing Practices

While the establishment of cGMP in Transfusion Medicine will improve transfusion safety through improving the control of operations, there may not be an accurate appreciation of the impact on cost and quality of care. Good Manufacturing Practices will reduce the number of mistakes and will stimulate more attention to quality management and quality engineering principles. These changes will protect against the risks which follow mistakes. Mistakes, however, are not thought to explain a significant portion of the remaining risk of infectious disease to which transfusion recipients are now exposed (Busch 1994). Therefore, the total impact on the safety of the blood supply will be significant but should not be exaggerated. Technology changes are needed to make a significant improvement in safety and will provide a more secure and rational platform to face future infectious disease risks. In addition, while cGMP has been very effective in drug and device manufacturing, conditions in Transfusion Medicine are not identical. Improvements in safety will be smaller than in classical drug synthesis or device manufacturing. Finally, implementing cGMP will increase the cost of Transfusion Medicine. The entire cost probably has not been felt yet, because the installation is not complete and because the fees currently charged may not cover current activity completely. These increased costs will be passed on to the end user and his/her insurance organization. Will these costs mean a 10% increase in fees levied? ... a 25% increase in fees? This significant increase will probably be in effect in many developed countries by the time red cell substitutes and viral inactivation are launched. It is not clear how increased costs will influence demand for new technologies.

The implementation of cGMP is mandatory in the United States and in companies abroad that wish to be licensed by the U.S. FDA. FDA Commissioner Kessler has said that eliminating the HIV window is a very high priority. Besides requiring cGMP, he has encouraged the pursuit of currently very expensive technologies to achieve this goal (Kessler 1994). Whether this same commitment to cGMP and further risk reduction will be made in all developed countries, especially those without a need for a U.S. license, remains to be determined.

1.3 Forces Affecting Transfusion Medicine

1.3.1 Trend Toward More Health Care

With the end of the Cold War, there may be less need for large defense budgets, and some resources previously consumed by defense might support improvements in health care. Most countries in the Western World show interest in reducing government expenditures and taxes. However, there has been an increase in attention to human rights and eliminating physical suffering, both within countries' borders and, when there is need, in neighboring countries. This new attention may lead to increased support for health care and Transfusion Medicine. Transfusion Medicine is a very visible portion of the health care system. The investigations that have been stimulated by concerns about the safety of the blood supply will bring more government interest and possibly resources to build stronger blood programs.

As part of its Global Program on AIDS, the World Health Organization (WHO) has had a Global Blood Safety Initiative. This initiative led to a new understanding of the needs and the level of service in many developing countries. On October 1, 1994, WHO made a commitment to establish a new unit dedicated to worldwide blood safety improvements (Emmanuel 1994). When this unit is established, WHO will attract talented professionals from around the world. It has the authority to communicate directly with governments and to bring resources specifically applicable to the particular circumstance found in each country. WHO could be extremely influential in building the foundation for dramatic improvements in health care worldwide by helping to place Transfusion Medicine programs in the proper order of priority and then helping to create unique paths to meet the specific needs of each developing country while controlling the use of resources.

The International Federation of Red Cross and Red Crescent Societies is the association that communicates with all the National Red Cross and Red Crescent Societies in the world. One mission of this program is to share resources and technologies so that National Societies' blood programs in developing countries can grow. The Red Cross and Red Crescent Blood Programs are non-governmental, but in many countries have been designated by the government to provide the needed services for the entire country. The Federation may strengthen its programs in the near future, and if this is accomplished, the Federation could bring improvements in Transfusion Medicine service to many developing countries (van Aken 1994). These improvements would speed the inevitable increase in demand for Transfusion Medicine services worldwide.

Finally, Europe has established the goal of self-sufficiency for blood and plasma derivatives and the standard that all blood and plasma derivatives will come from volunteer, non-remunerated altruistic donors (European Union Directive 1989). Japan has the same objectives. Currently the position is that countries should work toward these goals. It is the intention that eventually these would be requirements. The desire to generate internal resources and set a "higher" standard will stimulate Transfusion Medicine activity. Because much of the supply of plasma derivatives for Europe comes from the U.S. at present, and because a significant proportion of this U.S. supply is from paid donors, the accomplishment of the European and Japanese objectives would have worldwide implications for the plasma market.

1.3.2 Technology Improvements

1.3.2.1 Malignant Disease and Bone Marrow Transplantation

While there is no dramatic new therapy for malignant disease on the horizon, there are many advances that will decrease the risk of that therapy and have an impact on Transfusion Medicine. The intelligent application of growth factors and cytokines will speed the restoration of normal levels of blood cells after chemotherapy and could reduce the need for transfusion. At present there is no evidence that these substances can increase cure rates, increase survival or improve quality of life, but there is experience showing more rapid return of cells and possibly fewer infections. However, the observation that some tumor cells have receptors for growth factors is disturbing (Quesenberry 1994). An additional use of these substances is to stimulate the rapid growth of selected cells in culture. While cost and side effects are currently significant factors in the application of these substances, costs will probably decrease, and growth factors and cytokines will be applied in appropriate clinical and/or *in vitro* situations.

There has been much creativity in facilitating the collection and improving the quality of cells obtained for transplantation. The collection and transplantation of peripheral blood stem cells are realities today, and further growth will include allogeneic use of these cells. It is not clear what the potential impact of cord cells could be, but possible improvements might be less difficulty matching the recipient, more availability for autologous transplantation and possibly less Graft Versus Host Disease (GVHD).

Finally, improvement in the ability to purify and select cells for transplantation will support the prevention and/or the appropriate control of GVHD. GVHD may not remain the significant clinical problem it sometimes is today (Braine 1994). If GVHD can be prevented and controlled, hospitalization will be shorter, blood use will be less and costs will go down.

All these changes in technology will decrease the demand for blood components for each patient treated with standard or high-dose therapy or transplanted for malignancy. However, as the therapies become less risky and less expensive, the barriers for patients will be reduced and more patients will be recruited for this therapy. We have already seen an increase in the upper age limit for bone marrow transplantation in some centers. The prediction has been made that within the next ten years the number of transplants done will increase from about 7,000 worldwide to over 100,000 (Durbin 1988).

Currently, most of the blood transfused in developing countries is transfused as whole blood and goes to young children and women at the time of childbirth. In contrast, half the blood transfused in developed countries goes to patients over 60 years of age. This different pattern of usage suggests that malignant disease is not being treated aggressively in developing countries. As health care matures in developing systems, there will be an increase in the demand for the labile blood components. Improvement in health systems, the recruitment of patients and the aging of the population mean that the demand for transfusion will be significant.

1.3.2.2 Surgery

There has been a widespread effort among transfusion medicine specialists and clinicians to be more judicious in the use of blood components. The effort will increase and be supported by efforts to control cost and personal liability, but probably will not lead to a big decrease in demand for blood components.

A significant amount of blood is used in the resuscitation and definitive therapy of trauma patients. It has been learned that through aggressive attention during the first hour after trauma, many more patients are salvaged to receive definitive therapy. As more communities invest in support systems necessary to have an impact during this important first hour, there should be more patients surviving to have definitive surgery and rehabilitation. They will need the services of the Transfusion Medicine units.

1.3.2.3 Decontamination of Labile Cellular and Cell-Free Blood Components

Many health care manufacturing companies, biotechnology companies and nonprofit organizations are pursuing technologies that will further decrease the risk of transmitting infectious diseases with plasma derivatives in the short term and with cellular blood components in the longer term. Different technologies are being evaluated, including solvent detergent treatment, photoinactivation with and without various photoreactive chemicals, filtration, washing, other chemical reactions, *etc.* Currently, with the testing, screening and donor deferral systems in use, the risk of transmitting HIV with labile blood components in the U.S. is estimated to be in the range of 1:400,000/unit of blood; the risk of transmitting HBV, 1:200,000/unit; risk for HCV, 1:6,000/unit; risk for HTLV I/II, about 1:70,000/unit; risk for the rest, 1:1,000,000/unit or less (Dodd 1994). The risk of transmitting CMV to susceptible patients could be totally eliminated today simply by applying current filtration technology more aggressively. While these may seem like good odds to an individual who has not had or is not contemplating a transfusion, these seem to be frightening numbers to those with transfusion experience. The U.S. FDA Commissioner has said this level of risk is unacceptable (Kessler 1994). Vast resources are being invested worldwide to reduce and even eliminate the remaining risks. These investments will surely have some success. The time frame for the introduction of the new technologies is anybody's guess, and the temporal relation of the introduction of the decontamination technology to the introduction of red cell substitutes will be significant in determining the pricing and the preferred technology for various clinical indications.

1.4 The Arrival of Red Cell Substitutes

1.4.1 First Generation Red Cell Substitutes

We have, thus far, made some observations to help picture the blood banking world when the first red cell substitutes are launched. How will this world be changed by having a new material that transports O_2? Answering this question requires some assumptions about the first generation of substitutes. Assumption #1: More than one red cell substitute will be licensed in the first generation. Assumption #2: The first generation substitutes will have an *in vivo* half life ($T_{1/2}$) of fewer than 3 days. Assumption #3: Each red cell substitute will have different applications based on different characteristics.

One major appeal of red cell substitutes has been the ability to prepare them sterilely or to "sterilize" them once they are prepared. The public perception of the risk of transmitting AIDS has been the most significant stimulus for the investment in red cell substitutes. The solvent detergent procedure and the other processes used to decontaminate plasma derivatives have reduced the infectious disease risk of these substances well below the risk of cellular blood components. These processes will be applied to the modified hemoglobin solutions and such "sterile" substances will become the product of choice for the appropriate clinical indications. Recombinant technology will set the highest standard for reducing infectious disease risk. If, however, the technology to inactivate microorganisms in red cells has become available before the introduction of red cell substitutes, the immediate appeal of the substitutes will be reduced significantly.

In addition to the reduced infectious disease risk, cost, logistics of storage and administration, and new applications will be among the features that give red cell substitutes credibility in the marketplace. Storage will be more convenient because they are not likely to require refrigeration and they will be in a more concentrated form. They will not need to be crossmatched. These features will be advantages for the first generation of red cell substitutes. Disadvantages, when compared to red cells or whole blood, will be the short half life and the inevitable differences from normal physiology. If an improved perfluorocarbon is licensed, it may have some side effects different from those of hemoglobin solutions or cellular components. The balance of features will determine the demand for this first generation of products.

Because the $T_{1/2}$ will be short, the early substitutes will not be used for the same indications as red cells and whole blood. There will not be significant replacement of red cell transfusions. Because the viscosity of these early substances is likely to be lower than that of blood, additional uses will include perfusion distal to obstructions caused by plaques, catheters, *etc*. They may be used experimentally as support during shock and to determine if they improve salvage or reduce risk after massive trauma. They will be used to perfuse organs during storage before transplantation. The perfluorocarbon substitutes and substitutes based on synthetic hemoglobin will probably be acceptable for patients who refuse human blood transfusion or who are ineligible for human transfusion. This first generation of substances will be used widely in experimental situations to test new applications and to refine the true indications for red cell and whole blood transfusions.

It is difficult to predict the use of the first generation of red cell substitutes in developing countries. The red cell substitutes might be appropriate for bleeding at the time of childbirth to maintain these patients for a couple of days while hemorrhage is controlled and the marrow can replenish what has been lost. Young children, on the other hand, probably need a substance which will support them for some time. Expense will be a major issue in either case. Logistics of storage and administration should be very attractive.

This generation of materials will not have a great impact on demand for red cells and whole blood (see below). An effective perfluorocarbon would decrease the demand for donated human blood but to the extent that these substances have short half lives, they would still have limited ability to replace the bulk of the transfusions of red cells and whole blood.

1.4.2 Second Generation Red Cell Substitutes

One critical characteristic of future development in this area is the extension of the *in vivo* half life toward the half life of red cells transfused today. When this is achieved, we will see applications similar to the classic indications for today's red cell transfusions and there will be greater competition with components donated by volunteers. One way that the *in vivo* half life can be extended is through the encapsulation of the hemoglobin in lipid. This will accomplish a number of functions. It is likely to reduce concern about toxicity, because the encapsulated hemoglobin will prevent the exposure of all organs to hemoglobin with which they have little or no direct contact in normal physiology. Encapsulation will likely provide a longer *in vivo* half life. Encapsulation will also establish a more physiologic off-loading of O_2. While it is not clear that the timing of the transport of oxygen from hemoglobin to the tissue will be better if it has to include passage through a lipid membrane, this is what happens in normal metabolism. There are at least theoretical reasons why one would want to reproduce this situation with red cell substitutes. Viscosity could be less than blood, but probably greater than the hemoglobin solutions. It is probable that the first generation and the encapsulated products will exist in the market place concurrently, because the encapsulated substance may not be as useful for the perfusion indications listed for the soluble, non-encapsulated material.

It is when this longer half life material is licensed that significant change will occur in the blood centers of the world. If the longer half-life material is licensed before viral, protozoal and bacterial inactivation technologies are introduced, then the demand for the substitute will be very great. Usage will vary with cost, logistics of administration and storage and safety of the product. The company with the first substitute will establish pricing which will become the target for the other red cell substitute companies and for the viral inactivation technologies when they are introduced. If, on the other hand, treatment of red cells to reduce the risk of infectious disease transmission is routine when this longer half-life material is introduced, the demand will be less and will be more a function of cost and ease of use and storage. Whichever the order of introduction, treatment to reduce the risk of transmission of infectious disease is a competing technology for the red cell substitutes.

Eventually, red cell substitutes will contain hemoglobin made by recombinant technology. When this is accomplished, a new standard for infectious disease risk will be established. The recombinant red cell substitute will probably be considered to have the best risk/benefit ratio.

1.5 The Source of the Red Cell Substitute

The introduction of a red cell substitute will affect the activities of blood banks and blood bankers worldwide by reducing patient demand for classic blood components. As we have seen, it is likely that the half life of the first generation of materials licensed will be short and there may be only slight overlap in use and reduction in demand for donated blood components. When substances that reproduce physiology more precisely are licensed, they will replace current red cell containing components.

Another impact of the introduction of red cell substitutes will be on organizations providing the source material for the licensed product. If the material is human hemoglobin, then the impacts on blood centers would be different than if the material is synthetic or animal. If red cell substitutes do not require human red cells at any point in their manufacture, then the predominant indication for collecting human blood after the long half-life substitutes are approved would be for platelets and plasma proteins.

The red cell substitute can be synthesized from chemicals unrelated to hemoglobin as in the case of perfluorocarbons. Use of these substances would decrease the demand for human blood donations. If their indications are primarily new, then they may have little impact on the demand for human blood donations. It is less likely that a perfluorocarbon-based red cell substitute will reproduce normal physiology than a substitute based on hemoglobin will.

The red cell substitute could be hemoglobin-based but manufactured from animal red cells. The animal source (bovine or porcine, possibly) would be less expensive than other sources, but handling licensure, immunogenicity, allergy and the risk of infectious disease with these substances would provide new challenges for developers and regulators. Licensure of a red cell substitute containing animal hemoglobin would decrease the demand for human blood donations proportional to the degree to which the indication for their use matches the uses of classic red cell containing components.

Recombinant technology would be attractive because it could lead to the selection and production of precisely the right molecule and because the infectious disease risk might be smaller than that accompanying even the best technology which inactivates microorganisms. It is this feature which makes a recombinant source of hemoglobin the definitive source for the later generations of hemoglobin-based red cell substitutes. The development of synthetic molecules would likely be slower and more costly to accomplish in a cGMP fashion. Either recombinant or modified animal hemoglobin could be encapsulated. When a synthetic red cell substitute is licensed, there will be a reduction in demand for human whole blood donations both as source material and for direct transfusion. The impact on blood centers would be profound.

It is likely that for an interim period, the most desired red cell substitute might be based on human hemoglobin, and if that is the case, there would be two indications for the collection of whole blood donations, 1) to obtain red cells, platelets, plasma and whole blood for transfusion; 2) to obtain hemoglobin to manufacture red cell substitutes. Because the production of substitutes will undoubtedly lead to losses of hemoglobin, the production of red cell substitutes will require more donations than the production of an equivalent dose of red cells for transfusion. This interim period has already begun in one sense because of the high level of research and development in the field of red cell substitutes. Currently, much research and development work is being done with the hemoglobin obtained from outdated volunteer red cell and whole blood units, but it is not reasonable to think that outdated red cells would ever be sufficient if a product is licensed. The aggressive use of a human hemoglobin-based red cell substitute will create the demand for a new service, the collection of red cells for further manufacture. What policies would control this activity?

1.5.1 Red Cell Collection by Nonprofit Blood Centers

The greatest unchallenged claims for safety would exist if the source of the hemoglobin for further manufacture during this interim period was volunteer, altruistic, non-remunerated blood donors. The public and potential patient population are aware that volunteer blood is safer. Switching to paid donors might generate concern and would probably require defensive communication. While, with current processes to sequester, kill and remove microorganisms during fractionation, there is no evidence that volunteer plasma derivatives are safer than plasma derivatives based on source plasma from paid donors, in the future such evidence might become available. If that should happen, it would be appropriate to provide red cell substitutes manufactured from volunteer red cells. It is not the current practice, but regulatory agencies could make the volunteer source of red cells more attractive by requiring a source label as is done currently for cells and plasma for transfusion. The U.S. FDA requires that each blood bag carry a label that either says VOLUNTEER or PAID to indicate to the recipient the source of blood being transfused. The FDA does not require the same labeling of plasma derivatives because they are treated to reduce or eliminate the perceived difference in safety of the starting materials. It is possible that the FDA could establish this labeling standard for the red cell substitutes.

Finally, there is a clear perception among the governments and some blood banking professionals in Europe and Japan that volunteer plasma derivatives are/would be safer than plasma derivatives from paid donors. The result of the European (and Japanese) initiative to have all plasma derivatives come from volunteer, non-remunerated, altruistic donors is not clear (European Union Directive 1989). Some professionals believe that these countries will be successful in this effort and that by doing so, will establish a new world standard of safety for plasma derivatives. If that were to happen, it would be hard to imagine that the same standard would not apply to source hemoglobin for red cell substitutes. Others point out that there are currently no data to show that derivatives from volunteer sources are safer. They also say that the current European project would require dramatic changes in practice and investments of huge resources. They privately predict that Europe will never achieve its goal of meeting all needs with plasma derivatives from volunteers. Their prediction is that this movement will fade and will not require a volunteer source of hemoglobin for red cell substitutes.

To conclude this analysis, it might be appropriate to ask if the U.S. blood centers could possibly provide vastly increased quantities of blood to yield the hemoglobin for further manufacture if they were requested to do so. The blood centers would want to supply the hemoglobin. As the demand for their traditional services is eroded, providing hemoglobin would replace the lost activity and would be temporary protection for nonprofit centers. Currently, there is an intense internal focus in American blood banking. Many organizations are consumed with regulatory compliance and meeting internal and external quality standards. Finances are a significant problem in some communities. There has been a great deal of difficulty maintaining blood supplies adequate to meet the demands of current customers. It is, therefore, a stretch to accept that U.S. blood centers could, within the next five years, easily provide the source hemoglobin from volunteers even in the unlikely event that quality standards require it. The conditions in other parts of the developed world may be more conducive.

1.5.2 Red Cell Collection by the Manufacturers or by Contract

If the hemoglobin is not to come from nonprofit blood centers, from where would it come? The logistics of supplying the source hemoglobin from paid donors are probably simpler and the supply may be less costly as well. It costs less to pay donors than to recruit volunteers and maintain regular donations. Productivity in commercial source plasma centers is much higher than in the nonprofit collection centers for whole blood or for plasma and platelets. While the permitted frequency of plasma donation contributes to that productivity and "hemoglobin" donations could not be made as frequently, the source plasma centers would still have an advantage in providing low cost source material. There is automated technology for plasmapheresis that is attractive to the donors and maintains product quality. Similar technology might be made available for the collection of red cells and plasma simultaneously. In effect, if such technology were to be introduced, the fixed cost of the source plasma collection could be spread over a broader base of activity, which would generate greater revenue. Finally, the closer control of the source of the hemoglobin by the manufacturers and organizations with which they are used to dealing would remove the need to establish new, and probably difficult, relationships with nonprofit blood centers.

It is not clear whether source hemoglobin will come from volunteers or paid donors, but a paid source seems more probable. If, however, the hemoglobin comes from volunteer donors, then it is most likely that the collection would be done by the current nonprofit blood centers. It is said that volunteers will not freely give their blood to a profit making organization, because their gift would be turned into another person's profit. This old saw has never been tested sufficiently to be sure about its applicability to the current issue. Nevertheless, it is reasonable to assume that the requirement for a volunteer source of hemoglobin would mean that the nonprofits would have to collect it. If so, the new demand for red cells will provide a very significant challenge for the nonprofit blood centers. It will depend on the adoption of new automated technologies. The successful program will require major change in their management and culture.

If on the other hand, the supply of hemoglobin is provided by the source plasma centers, the impact on their operation will be significant but far smaller. If the hemoglobin comes from paid sources, as the applications for the red cell substitutes grow, the nonprofit blood centers will be requested to collect fewer units of blood. In addition, the licensure of a long half-life red cell substitute would decrease the demand for current red cell products for transfusion and force further downsizing of the nonprofits. They will be required to reconfigure the collections operations to emphasize platelets and plasma for transfusion. Finally, when recombinant technology provides the hemoglobin in the substitute, the blood and source hemoglobin/plasma centers will only be in the business of providing platelets and plasma. It is not inconceivable that the productivity and the safety of the red cell substitutes and the plasma derivatives might push the source plasma/red cell centers into the collection of platelets for transfusion as well. This latter possibility would become more likely if there were an effective platelet viral inactivation procedure.

1.6 Impact on Developing Countries

When viewing the progress of technology in developing countries, it is tempting to picture the reproduction of the technologies and the procedures that are so successful in the developed countries. Indeed, that is likely to be the result sought by those helping the developing countries and by the professionals in those countries as well. One might conclude that if the procedures and technologies work in the U.S. or Japan, they would have to be best for a developing country as well. This then becomes the path of least resistance for the source of the aid, and it is a path which is very much desired by the professionals in the developing country. The issue may not be so simple and deserves a second look.

1.6.1 The Situation in Each Country is Unique

The infectious disease challenges faced in each country will not be the same. And it is not just a matter of having to do different tests. In some countries the frequency of infection among "healthy" donors is so high as to make testing very unattractive. Sorting the undesirable blood and preventing its use would be a challenge and, of course, the number of tests per "desirable" unit would be very high. In addition, the system to keep track of donor test information is required with current enzyme-linked immunoassays and confirmatory tests because the result of a previous test influences the handling of subsequent donated units. This is not only a significant expense but would provide a logistic challenge in many places. Once infected donors are discovered, a major burden for their welfare is placed on the Transfusion Medicine resource that could prevent appropriate growth. This present system is not completely satisfactory even in the U.S., and it is very costly. Many professionals want an opportunity to find a new system to provide safe blood.

Developing countries will have limited resources to start and maintain blood transfusion systems. If their resources cannot support the establishment of multi-tiered safeguards, which are in fashion in the West at present, these countries must still make an investment to achieve the maximum quality possible with available resources. If they set up systems designed for their specific needs, they may just achieve adequate safety at a much lower cost.

Given the conditions in underdeveloped countries, how will the emerging technologies provide them opportunities to implement safe transfusion systems in support of the rest of health care? Will viral inactivation and red cell substitutes have a place? Will these countries use these technologies in the same fashion that they are used in the West? The answers to these questions should be different in each country. In the U.S., there is likely to be a period when blood is obtained from thoroughly tested and screened volunteers and then decontaminated to remove all microorganisms. In developing countries, it is possible that viral inactivation would be used without as much investment in donor or blood screening. It is possible that red cell substitutes will be applied in more clinical situations than in developed countries because of the ease of storage and administration.

When systems to inactivate a broad range of infectious microorganisms are mature, these might be especially attractive to countries that have a high frequency of carriers for diseases transmissible by blood. Technology to kill infectious agents would be more productive than technology to test for these organisms, because the frequency of unacceptable units will be high. This technology could support a safe blood program before public health efforts eradicate endemic diseases transmitted by blood.

Red cell substitutes will also be evaluated differently in developing countries. Their use obviously will depend on the needs. During the period when only short half-life substitutes are available, use might be limited. Once the half life of the red cell substitutes increases, the substitutes could be the most logistically feasible way to support the health care system in some communities. If the developing countries can afford the substitutes, they will probably have wider applications than in the developed countries.

In this chapter we cannot evaluate the circumstances of every developing country and predict how the new technologies will influence the improvements in health care in each. What is important, however, is to note that new technologies increase the probability of building a system designed for the circumstances of each country. This may take coordination of the developers of technologies and health service agencies that are responsible for helping transfusion services of developing countries. The need in these countries is great and until there is the possibility of "good business," it is not likely that the appropriate resources can be coordinated. Once that coordination is occurring, lives will be saved.

1.7 Summary and Conclusions

The critical message for health care planners today is that there is probably more potential change in Transfusion Medicine than can be followed and more than was present at any other time in the history of Transfusion Medicine. This change is primarily in the technologies used within the specialty and in response to changes elsewhere in health care. Safety and cost are the prime drivers. Viral inactivation and red cell substitutes are being pursued, at least partially, because of a public perception that blood is not as safe as it should be. The public's dissatisfaction is not based on objective assessment or comparison to an accepted standard for the safety of blood components. While there is much room for improvement in the safety and efficacy of blood components, they are safer than the public thinks. The new technologies will be more reassuring to the public and provide more protection against known and un-known microorganisms. The new platform will be better because blood components will be designed to accomplish a specific purpose. Effects other than the desired ones will be much better controlled. Possible con-taminants will be more effectively eliminated. Infectious agents can be kept out of the components during manufacture. This is far better than testing and sorting to remove the unacceptable units. But the actual de-crease in risk will be small, and the odds of getting a disease from a blood component in a developed country will go from low to slightly lower.

The stresses on blood center and system managers trying to respond to weaknesses are great. They will increase because of health care reform, attention to cost, competition and new technologies. Over the next years, significant forces will challenge the currently stable condition of non-profit regional blood centers.

The technologies on the horizon have a great potential to improve the situation of today's transfusion recipients. One of the biggest potential impacts is in the developing countries. While there is still a serious shortage of resources in most of these countries, without these new tech-nologies, there may be no cost-effective way to proceed. Once red cell substitutes and blood decontamination are available, there will be clearer paths to accomplishing the goals of safe, adequate and effective blood supplies. As health care is improved and more resources are allo-cated in these areas, the potential technologies may be introduced suc-cessfully in these countries with an enormous relative increase in the safety of the recipient of blood components.

1.8 References

Braine, H. Professor, Oncology and Medicine, The Johns Hopkins Uni-versity School of Medicine, Baltimore, MD, Personal Communication, 1994.

Brand, A.. Passenger leukocytes cytokines, and transfusion reactions. *N. Eng. J. Med.* 331: 670-671, 1994.

Busch, M. Scientific Director, Irwin Memorial Blood Center, San Fran-cisco, CA. Remarks at the FDA Conference, *The Feasibility of Genetic Technology to Close the HIV Window in Donor Screening,* Silver Spring, MD, September 26, 1994.

Chassaigne, M. Directeur, Centre Regional de Transfusion Sanguine, Tours, 2, Boulevard Tonnelle (CHRU) BP 2009 37020 Tours, Personal Communication, 1994.

Dodd, R. Head, Transmissible Disease Department, Holland Labora-tory, American Red Cross Biomedical Services, Remarks at the FDA

Conference, *The Feasibility of Genetic Technology to Close the HIV Window in Donor Screening*, Silver Spring, MD, September 26, 1994.

Durbin, M. Bone marrow transplantation: economic, ethical and social issues. *Pediatrics*, 82: 774-783, 1988.

Emmanuel, J., Acting Chief, Global Blood Safety Unit, World Health Organization, Geneva Switzerland, Personal Communication, 1994.

European Union Directive 89/381/EEC

Kessler, D. Commissioner, U.S. Food and Drug Administration, Remarks to the FDA Conference, *The Feasibility of Genetic Technology to Close the HIV Window in Donor Screening*, Silver Spring, MD, September 26, 1994.

Quesenberry, P.J., H. Ramshaw, S. Peters, S. Rao, C. Tiarks, P. Lowry, M. Stewart, P. Becker, P. Newburger, and G. Stein. Normal hemopoiesis. In *Blood Supply: Risks, Perceptions, and Prospects for the Future* (S.T. Nance, ed.), Bethesda: American Association of Blood Banks, 1994.

van Aken, W.G., Director, Central Laboratory of the Netherlands, Red Cross Blood Transfusion Service, Amsterdam, The Netherlands, Personal Communication, 1994.

Chapter 2

Demonstration of the Efficacy of a Therapeutic Agent

Joseph C. Fratantoni, M.D.

Division of Hematology, Center for Biologics Evaluation and Research, Food and Drug Administration, 1401 Rockville Pike (HFM330), Bethesda, Maryland 20852-1448

2.1 Introduction

The FDA is charged with ensuring that drugs and biologics are pure, potent, safe and effective. In this brief discussion, we will consider the regulatory basis for the demonstration of clinical efficacy required by FDA. We will also discuss the nature and variety of clinical trial endpoints.

2.2 Federal Regulations

The Code of Federal Regulations speaks directly to the issue:

> "Effectiveness means a reasonable expectation that, in a significant proportion of the target population, the pharmacological or other effect of the biological product ... will serve a <u>clinically significant function</u> in the diagnosis, cure mitigation, treatment or prevention of disease in man." (21 CFR 601.25(d)(2))

We need to pay special attention to the term "clinically significant function", which I have underlined for emphasis. This means that the agent under evaluation must be shown to clinically benefit the patient population that is being studied. Put another way, the primary endpoint of the pivotal clinical trial that is intended to support approval must be a direct measurement of the clinical benefit of the agent. There are a limited number of types of endpoints that satisfy the above definition: 1) increase survival of the study population; 2) provide measurable sympto-

Blood Substitutes: Physiological Basis of Efficacy
Winslow et al., Editors
© Birkhäuser Boston 1995

matic relief to the study population; 3) prevent or slow the progression of disease. In many cases, it is not possible to make measurements of such direct endpoints, and investigators seek to substitute a more readily measured entity as a surrogate for a clinical endpoint. The authority to use such an alternative is also contained in the Code of Federal Regulations:

> "Proof of effectiveness shall consist of controlled clinical investigations ... unless this requirement is waived on the basis of showing that it is not reasonably applicable to the biological product or essential to the validity of the investigation and that an alternative method of investigation is adequate to substantiate effectiveness. Alternative methods ... may be adequate ... where a previously accepted correlation between data generated in this way and clinical effectiveness already exists." (21 CFR 601.25(d)(4))

2.3 Surrogate Endpoints

Such a surrogate endpoint of a clinical trial is a laboratory measurement or physical sign that is a substitute for, and is expected to correlate with, a clinically meaningful endpoint that directly measures how a patient feels, functions or survives. Hypertension is a surrogate marker for hypertensive cardiovascular disease, an example we shall return to later.

There are two principal risks in using a surrogate endpoint: 1) the clinically meaningful endpoint may not actually correlate with the proposed surrogate, even though it was thought to do so. In this case, adverse consequences of the drug are a clear net loss; 2) the surrogate may, in fact, correspond to a real benefit, but the drug may have serious undesirable consequences as well as the effect on the surrogate, complicating evaluation of the true risk:benefit ratio for the agent under investigation.

In general, surrogate endpoints can be considered when there is sufficient knowledge of the disease treated and the agent under investigation, when the feasibility of performing meaningful clinical trials is poor, and when the overall risk:benefit situation justifies such use.

A few specific examples of surrogate clinical trial endpoints will help to illustrate the above concepts. Hypertension is known, by virtue of extensive clinical studies and experience, to be a causative factor in cardiac and cerebral vascular disease. Similarly, measures that decrease the blood pressure of a hypertensive population have clearly been shown to decrease the incidence of those diseases. Agents which lower blood pressure have been accepted as effective agents for prevention of cardiac and cerebral vascular disease and, therefore, to have a clinical benefit. This correlation is based on extensive clinical and experimental data.

Coagulation factors intended for use in treating patients with hemophilia have been used for many years. In the case of hemophilia A, a therapeutic factor concentrate that increases the *in vitro* assay for factor VIII will also increase the factor VIII level when infused into a patient, and extensive studies have shown that this set of circumstances will permit that patient to withstand more severe hemostatic challenges (*e.g.*, surgery, dental extraction) than if the infusion had not been given. Accordingly, coagulation-factor products prepared using modified procedures (*e.g.*, more extensive viral inactivation measures) have been approved on the basis of *in vitro* demonstrations and pharmacodynamic studies in patients. An interesting addendum to this example is the observation that, in the case of the new recombinant factor VIII preparations, additional studies were needed, since the agent was not as well known, and other concerns were raised (such as possible increased antibody formation).

A final example of the use of surrogate endpoints may be obtained by examining the approval of recombinant erythropoietin (r-EPO). This agent was studied in patients with chronic renal failure, most of whom required transfusions to maintain their hemoglobin levels at a suitable level. The endpoints accepted for approval were the achievement and maintenance of a target hemoglobin level and a decrease in the need for allogeneic transfusions. Since EPO stimulates the formation of red cells, and since the beneficial effect of red cell transfusions was well known from extensive clinical experience, the use of a hemoglobin level as a surrogate endpoint is understandable. Because of the concerns about the unwanted infectious and immunologic adverse reactions to allogeneic transfusion, decreasing the need for such transfusion is a valid surrogate endpoint.

There are other situations that exemplify the limits of surrogate endpoints, such as the CAST study (Cardiac Arrhythmia Suppression Trial). This study involved drugs that suppressed extrasystolic ventricular contractions. We know that people with these arrhythmias are at increased risk of death following a heart attack, so intuitively it seemed that a reduction in irregular heartbeats would predict reduced mortality. In fact, the drugs that reduced arrhythmias were associated with a 250% increase in mortality, related to other effects of the agents used.

Sometimes, the opposite scenario unfolds, as was the case with the use of gamma interferon for chronic granulomatous disease (CGD). Investigators expected the agent to affect superoxide production and thereby render a favorable effect on CGD patients. The trial was begun, and although there was no effect on superoxide, the patients showed clear clinical improvement, a phenomenon yet to be fully explained (International Chronic Granulomatous Disease Cooperative Study Group 1991).

2.4 Efficacy *versus* Activity

We should complete this discussion by again referring to two elements of clinical studies that are often confused:

- Efficacy is demonstrated by production of a clinical benefit.

- Activity is demonstrated by results obtained in a biological or chemical or physical assay.

In the evaluation of red cell substitutes, an increased oxygen tension or a higher hemoglobin level after infusion of a hemoglobin solution would be demonstrations of chemical or biologic activity. Improved function of an organ as the result of perfusion with a test material could be a measure of efficacy, depending upon the context of the study.

In order to facilitate the design of red cell substitute studies, FDA proposed stratification of potential indications, *e.g.*, local perfusion, perioperative use and hemorrhagic shock (Center for Biologics Evaluation and Research 1994). Improved organ function has been used for approval of Fluosol in local perfusion of the coronary arteries during angioplasty. Clinical benefit in hemorrhagic shock will probably require demonstration of decreased mortality as related to the particular circumstances involved. For perioperative use, one might consider that the standards of current medical practice consider allogeneic red cell transfusion an outcome to be avoided. Therefore, if use of a test agent can be shown to decrease dependence on allogeneic transfusion, the agent would provide a clinical benefit and would be considered "effective".

2.5 Summary

In summary, demonstration of the efficacy of red cell substitutes must follow the same approaches required for other drugs and biologics. The situation with red cell substitutes is complicated by several factors: 1) the biological behavior of the test agents is not well understood; 2) the basis for the efficacy of red cell transfusion, while accepted, is not readily explained in many instances; 3) red cell transfusions are quite safe, despite contrary public perception, thereby making a satisfactory risk:benefit state more difficult to attain. While there are some unique difficulties associated with demonstrating the efficacy of red cell substitutes, the intense interest in the field and the rapidly increasing understanding of the underlying biology support the contention that this effort will ultimately be successful.

2.6 References

Center for Biologics Evaluation and Research Points to Consider on efficacy evaluation of hemoglobin- and perfluorocarbon-based oxygen carriers. *Transfusion* 34: 712-713, 1994.

International Chronic Granulomatous Disease Cooperative Study Group. A phase III study establishing efficacy of recombinant human interferon gamma for infection prophylaxis in chronic granulomatous disease. *N. Engl. J. Med.* 324: 509-516, 1991.

Chapter 3

A Physiologic Basis for the Transfusion Trigger

Robert M. Winslow, M.D.

Department of Medicine, School of Medicine, University of California, San Diego, Veterans Affairs Medical Center (111-E), 3350 La Jolla Village Drive, San Diego, California 92161

ABSTRACT

Allogeneic blood for transfusion is now rigorously tested and the risk of contamination with pathogens is extremely low. The decision to transfuse a given patient should be based on the need for transfusion, not just on an estimation of the risks. Transfusion requirements need to be understood on physiological grounds, but there is no clear-cut "transfusion trigger" that emerges from this understanding. Rather, the clinician must combine physiologic parameters such as $P\bar{v}O_2$, $S\bar{v}O_2$, O_2 extraction ratio and $\dot{V}O_2$ (when available) with clinical signs and symptoms to provide a rationale for transfusion of individual patients.

3.1 Introduction

For many years, if a preoperative patient's hemoglobin was less than 10 g/dl he received at least 2 units of packed red cells or whole blood, in spite of the well known fact that many patients tolerate modest anemia quite well (Mollison 1983). The rationale for such transfusions was that an O_2 reserve needed to be maintained so that if unexpected (or expected) blood loss occurred during surgery, the patient would be in less danger of suffering deficient O_2 delivery to tissue.

In the early 1980's the medical and lay communities became acutely aware of the infectious risks of blood transfusion. The possibilities of mismatch or contaminated units have always been appreciated, but the new specter of AIDS transmission by banked blood forced critical attention to the indications for red cell transfusion. Today, these risks, together with the growing economic concerns over optimal use of blood and

Blood Substitutes: Physiological Basis of Efficacy
Winslow et al., Editors
© Birkhäuser Boston 1995

blood products, are sharpening our attention to the indications for red cell transfusion.

Because of extensive quality control testing, the nation's red cell supply is safer now than it ever has been in the past (Dodd 1992). But, as pointed out in the chapter by Tomasulo (1995), it is impossible to specify all of the components in a unit of red blood cells, and the likelihood that such specification will ever be possible is remote. Thus, we cannot deny that new, as yet undiscovered, dangerous infectious agents *might* be transmitted by blood transfusion, and we are obliged to endeavor, as clinicians, to transfuse only when there is a clear clinical indication.

The "transfusion trigger" is that event or set of events which results in a patient's receiving a red cell transfusion. Excellent recent discussions have been published regarding the transfusion trigger (Stehling and Simon 1994, American College of Physicians 1992, Goodnough *et al.* 1992, Levine *et al.* 1990), and numerous conferences have been held to attempt to set up specific guidelines or algorithms whereby clinicians can make objective decisions regarding the use of red cells (NIH Consensus Conference 1988). Nevertheless, there remains no such guideline which would reliably serve a clinician who is not familiar with the basic principles of O_2 transport physiology.

It is my contention that even if blood were entirely safe in terms of transmission of infectious agents, good medical practice would dictate that red cells are transfused only when there is a good clinical indication. This implies an understanding of the basic principles of O_2 transport physiology as it applies to red cell transfusions, and the purpose of this chapter is to review these principles.

3.2 Oxygen Transport: Definition of Terms

Global (sometimes called *convective*, as opposed to capillary or *diffusive*) O_2 transport is the product of the blood flow (cardiac output) and the difference between arterial and mixed venous blood O_2 content. This relationship is simply a statement of the well-known Fick equation:

$$(1) \qquad \dot{V}O_2 = \dot{Q} \times (CaO_2 - C\bar{v}O_2)$$

Where $\dot{V}O_2$ is the O_2 utilized by the body, $\dot{Q}$ is the cardiac output, CaO_2 and $C\bar{v}O_2$ are the arterial and mixed venous O_2 contents, respectively.

Diffusive O_2 uptake in the lung (see Figure 3.1) is described by the diffusion equation:

$$(2) \qquad \frac{d(O_2)}{dt} = \frac{100}{Vc} \times \frac{DLO_2}{60} \times (PAO_2 - PcO_2)$$

In this equation, $d(O_2)/dt$ is the rate of O_2 diffusion into the capillary, PAO_2 is the alveolar PO_2, PcO_2 is the pulmonary capillary PO_2, DLO_2 is the diffusion coefficient for tissue and Vc is the volume of capillary blood. Roughton and Forster (1957) described the components of this coefficient:

$$(3) \quad \frac{1}{DLO_2} = \frac{1}{DMO_2} + \frac{1}{\theta O_2 \times Vc}$$

In this equation, DMO_2 is the diffusion coefficient for the alveolar/capillary membrane interface and is proportional to the thickness of the membrane. θO_2 is the off-reaction rate constant of O_2 with hemoglobin.

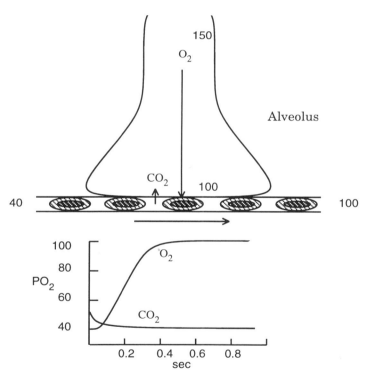

Figure 3.1 Diffusive uptake of O_2 and release of CO_2 in the lung. As venous red blood cells pass into pulmonary capillaries, they take up O_2 from plasma which is oxygenated by diffusion across the pulmonary capillary membrane. The average time of transit for a red cell under resting conditions is approximately 0.8 seconds.

These relationships are conceptually simple, but their numerical modeling is complex and iterative. For example, θO_2 is dependent on the hemoglobin saturation, which in turn, is dependent on the position of the O_2-hemoglobin saturation curve (P50), itself dependent on 2,3-DPG, pH, and PCO_2. pH and PCO_2 are further interdependent, and both are affected by the buffering of hemoglobin. Each of the terms in equations 1-3 has a number of determinants, which can vary in different physiological conditions and in disease or pathological states. While complex, all of these variables can be accounted for using computer techniques, and, as shown in Figure 3.1, exchange curves for O_2 and CO_2 in the lung can be calculated (Winslow 1985). The situation in the tissues is slightly more complicated, but in general represents the mirror image, and the principles governing the exchange of gases are the same. Another, perhaps clearer, way to view the Fick relationship is shown graphically in Figure 3.2. Each of the panels of the figure is made up of two components: the hemoglobin-O_2 dissociation curve and the cardiac output. The numerical calculations, summarized in Table 3.1, for the three examples to follow were done based on data from patients and normal humans studied under different physiological conditions (Winslow 1985).

3.3 Examples

The examples to be discussed consider a number of variables which can alter overall O_2 transport individually. In the calculations, the cardiac output is determined by 2 factors: the O_2 requirement ($\dot{V}O_2$) and the blood viscosity (determined by the hematocrit). The regulation of these variables is well documented in the literature (Guyton, Jones and Coleman 1973). The mixed venous PO_2 ($P\bar{v}O_2$) is calculated as that PO_2 which corresponds to the mixed venous O_2 content when all conditions in the Fick equation are satisfied. These calculations and the assumptions underlying them have been described previously in the literature (Winslow 1985). The variables considered in these examples are the following:

1. Alveolar PO_2 (PAO_2). This is assumed to be 100 torr, corresponding to a normal sea level environment in a subject with normal ventilation.

2. Arterial PO_2 (PaO_2). This is also assumed to be 100 torr, and assumes no alveolar-arterial diffusion gradient. In fact, a small gradient usually is present, but does not affect the calculations because the blood is well-saturated even at 80 torr.

3. Arterial PCO_2 ($PaCO_2$). This is assumed to be 40 torr, the value found in normal resting persons with normal ventilation.

4. Arterial pH is assumed to be 7.4, a normal value.

5. 2,3-DPG. This is actually the red cell 2,3-DPG/hemoglobin molar ratio. In a large number of normal volunteers studied in our laboratory, this value is 0.88 mole/mole (Samaja and Winslow 1979). The 2,3-DPG/Hb ratio, pH, and PCO_2 are the principle determinants of the position of the blood O_2-dissociation curve (P50). These parameters are used

to calculate a continuous curve for the given conditions used in the example calculations (Winslow *et al.* 1983).

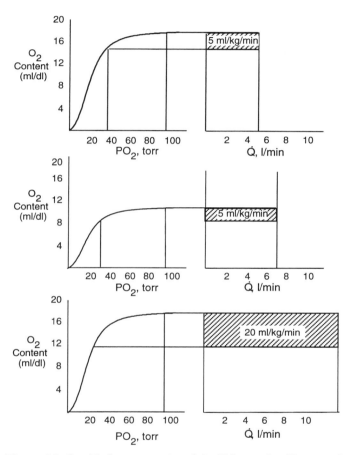

Figure 3.2 Graphical representation of the Fick equation. The examples are described in the text and include normal resting conditions (top), anemia (middle), and increased O_2 demand (bottom) as in mild to moderate exercise. A numerical summary is provided in Table 3.1.

6. Pulmonary membrane diffusion constant (DMO_2). This is assumed to be 40 ml/ml/mm Hg, as it is in normal persons. Alterations of this value will change the alveolar-arterial diffusion gradient (Wagner 1977).

7. The volume of pulmonary capillary blood (Vc). This volume also is a parameter of pulmonary diffusion (Roughton and Forster 1957) and in certain pulmonary diseases can increase the alveolar-arterial diffusion gradient. A value of 70 ml is used.

3.3.1 Normal Hematocrit and O_2 Demand

The first example is a normal resting human with hematocrit 45%. In this example, the O_2 utilization is approximately 5 ml/kg/min and the arterial O_2 content is about 18 ml/100 ml of blood. To satisfy the conditions that there is no base excess, that the O_2 dissociation curve is determined by the factors mentioned above, and that the cardiac output (as determined by hematocrit and $\dot{V}O_2$) is 5.46 l/min, the arterial-venous O_2 content difference results in a $P\bar{v}O_2$ of 37.4 torr. These calculations show that 34% of the arterial O_2 is extracted by tissue.

3.3.2 Anemia, Normal O_2 Demand

The second example considers holding all variables constant except that the hematocrit is dropped to 25%. As a result of this change, the cardiac output rises to 7.42 l/min by virtue of reduced blood viscosity and resistance to flow. This increased flow, however, does not completely compensate for the reduced arterial O_2 content, and a larger fraction of the arterial O_2 must be extracted (44%), resulting in a lower $P\bar{v}O_2$ (32.0 torr). Still, however, the O_2 requirement (5 ml/kg/min) can be satisfied.

Table 3.1 Oxygen exchange parameters for example cases in Figure 3.1.

	$P\bar{v}O_2$	$\dot{Q}$	OER
Hct = 45%	37.4	5.46	0.34
Hct = 25%	32.0	7.42	0.44
$\dot{V}O_2$ = 20ml/kg	27.9	13.3	0.53

PAO_2 = 100 torr, PaO_2 = 100 torr, $PaCO_2$ = 40 torr, arterial pH = 7.4, 2,3-DPG/hemoglobin = 0.88, DMO_2 = 40 ml/ml/mm Hg, Vc = 70 ml. OER = oxygen extraction ratio.

It should be appreciated that this example is a simplification, since in anemia, ventilation increases, raising PAO_2, lowering $PaCO_2$ and raising pH. In turn, the changes shift the O_2 dissociation curve to the left (increased affinity, lower P50) which, in turn, affects θO_2. Furthermore, with time, this respiratory alkalosis becomes compensated by metabolic (renal) mechanisms.

3.3.3 Normal Hematocrit, Increased O_2 Demand

The third example illustrates still another variation on these variables: increased $\dot{V}O_2$, as might be found in moderate (but aerobic) exercise. In this case, the $\dot{V}O_2$ is placed at 20 ml/kg, but all other variables are held constant. The result is that the cardiac output is much higher (13.3 l/min), the mixed venous PO_2 is lower (27.9 torr), but a larger fraction of the arterial O_2 must be extracted (53%). But again, all conditions are satisfied.

This, too, is a simplification of the actual situation: hyperventilation again may increase PAO_2 and DMO_2 with resulting alkalosis, but muscular work will also produce lactic acid which drops pH with the effect of reducing O_2 affinity (increased P50). Furthermore, raised cardiac output shortens the "dwell" time for a red cell in the pulmonary capillary and when raised to extreme values, can actually lead to reduced PaO_2 during heavy exercise (see Figure 3.1).

3.4 Physiological Transfusion Triggers

Using this brief quantitative framework, is it possible to identify physiological markers that can be used as a transfusion trigger which might be more useful than hemoglobin or hematocrit? Obviously, a number would be more useful in clinical practice than subjective measurements and could lead to uniform criteria for the transfusion of red blood cells.

As our examples show, when the hemoglobin drops, as the demand for O_2 rises, more of the arterial O_2 is used, venous O_2 is depleted, and $P\bar{v}O_2$ falls. Our calculations can be used to explore a range of O_2 delivery values ($\dot{Q}$ x CaO_2 or DO_2) to predict what effect might be seen on $P\bar{v}O_2$ (Figure 3.3). In this figure, the DO_2 was reduced by reducing hematocrit at 5% intervals and calculating various quantities shown in Table 3.1.

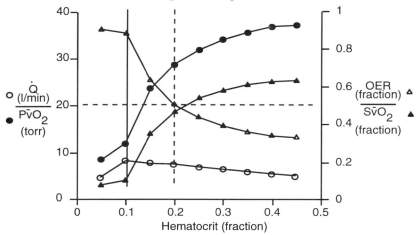

Figure 3.3. The effect of decreasing O_2 delivery (hematocrit) on potential transfusion triggers. A rising cardiac output and falling mixed venous PO_2 ($P\bar{v}O_2$) and saturation ($S\bar{v}O_2$) compensate over most of the range. All parameters appear to be continuous functions and do not provide clear "triggers". The horizontal dashed line shows an OER of 50%. If this were used as a transfusion trigger, transfusion would be given at a hematocrit of 20%.

As hematocrit falls, $P\bar{v}O_2$ drops and the oxygen extraction ratio rises as continuous functions over the range of anemia down to a hematocrit of about 10%. While we would agree that a transfusion would be desirable at some point in that interval, the data do not provide a clear-cut transfusion trigger.

The relationship between O_2 supply and utilization is shown in Figure 3.4. With $\dot{V}O_2$ set at 5 ml/kg/min, the hematocrit was dropped in 5% steps, and the parameters shown in Table 3.1 were calculated. Cardiac output increases, $P\bar{v}O_2$ decreases, and the oxygen extraction ratio becomes greater, as shown in Figure 3.3. But a point occurs - here at a hematocrit of 10% - when these normal compensations no longer serve, and the only way to complete the calculation is to decrease $\dot{V}O_2$. This point is labeled the "critical DO_2".

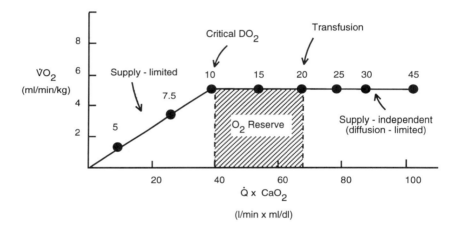

Figure 3.4. The critical DO_2 ($\dot{Q}$ x CaO_2). As the hematocrit is reduced in 5% steps, $\dot{V}O_2$ is maintained by a combination of rising cardiac output and dropping $P\bar{v}O_2$ (see Figure 3.3). When these compensations no longer suffice, $\dot{V}O_2$ falls, and tissue ischemia occurs. The goal of transfusion therapy is to maintain the O_2 reserve such that this point is not reached.

The critical DO_2 is the point at which tissue O_2 delivery can no longer be satisfied, and tissues become hypoxic. Some produce lactic acid, but others suffer compromised function. Obviously, the point is to be avoided in clinical practice, but as shown in Figure 3.3, there is no clear indication from easily measured quantities what that point might be. Thus, the goal of transfusion is to maintain an "O_2 reserve" such that the critical DO_2 is not reached.

3.5 Physiologic Triggers

This analysis does not provide a physiologic transfusion trigger. Many parameters have been proposed as indicative of the need for transfusion, but none, by itself, is sufficient. The thoughtful clinician observes all possible signs of tissue ischemia and, based on experience, attempts to transfuse before the critical DO_2 is reached. In the following section, a brief discussion of some of the useful potential triggers will be discussed briefly.

3.5.1 Mixed Venous PO₂

Mixed venous PO_2 ($P\bar{v}O_2$) would seem an obviously important parameter to assess adequacy of tissue oxygenation since, in theory, the mixed venous blood should be in equilibrium with tissue. However, as is now known, the tissue PO_2 is much lower than mixed venous values, and $P\bar{v}O_2$ can be normal in severe anemia (Gould *et al.* 1983). The reason for this is complex but includes the fact that capillary blood can be shunted to different vascular beds in the face of hypovolemia, for example, so that $P\bar{v}O_2$ does not necessarily reflect the oxygenation state of <u>all</u> tissues (Messmer *et al.* 1972). In addition, it is now understood that there is significant O_2 loss from arterioles and uptake in venules, so that the capillary PO_2 is lower than mixed venous values.

Nevertheless, a decreasing $P\bar{v}O_2$ has been used as a classic indicator of reduced tissue oxygenation, and perhaps a dropping value should be more meaningful to clinical evaluation of a given patient than an absolute value. Traditional textbooks of critical care medicine indicate that transfusions may be helpful when the $P\bar{v}O_2$ is less than 20 torr.

Mixed venous oxygen saturation ($S\bar{v}O_2$) may be a more useful indicator of severe O_2 extraction. Because of the steepness of the hemoglobin-O_2 dissociation curve, when mixed venous PO_2 falls below approximately 30 torr, the hemoglobin saturation falls rapidly (see Figure 3.3). In one clinical study, Spiess *et al.* (1992) showed that $S\bar{v}O_2$ is a sensitive indicator of overall O_2 consumption in liver transplant patients. In that case, removal of the patient's liver with attendant reduction in $\dot{V}O_2$ produced a measurable rise in $S\bar{v}O_2$. When the new liver became functional, $S\bar{v}O_2$ fell back to normal values.

Unfortunately both $P\bar{v}O_2$ and $S\bar{v}O_2$ can only be measured in pulmonary artery blood by use of an indwelling catheter.

3.5.2 O₂ Consumption

Published literature suggests that reduced $\dot{V}O_2$ in postoperative and trauma patients is associated with a poor prognosis, and that increasing DO_2 by intervention (fluid boluses, administration of blood products, the use of inotropes) reduces mortality rate (Shoemaker *et al.* 1985). In contrast, other evidence indicates that the *way* in which DO_2 is increased is critical: volume resuscitation more effectively raises $\dot{V}O_2$ than transfusion, even though both raise DO_2.

In septic shock, $\dot{V}O_2$ may be pathologically dependent on DO_2 (see Figure 3.4) (Ronco *et al.* 1990). Patients whose cardiac output (and therefore DO_2) can be increased with dobutamine (Mink and Pollack 1990; Lorente *et al.* 1993) or adrenalin (Seear, Wensley and MacNab 1993) can increase $\dot{V}O_2$. However, when DO_2 is raised by increasing hemoglobin concentration by transfusion, no effect is seen on $\dot{V}O_2$ (Hanique *et al.* 1994; Seear Wensley and MacNab 1993 Lucking *et al.* 1990). In fact, when 23 critically ill patients with sepsis were transfused with stored blood, not only did the $\dot{V}O_2$ (measured by calorimetry) fail to rise, but there was an inverse relationship between gastric mucosal pH and the age of the transfused blood (Marik and Sibbald 1993), indicating poorer tissue oxygenation. These studies all indicate that pathological reduction of $\dot{V}O_2$ in sepsis is due to reduced tissue perfusion, not reduced O_2 content of the perfusing blood.

A part of the problem in rationalizing these data may be that any calculated $\dot{V}O_2$ based on the Fick equation (see equation 1) depends on the patient or animal being in the steady state. The steady state, briefly, is that condition in which the O_2 taken up exactly equals the O_2 being utilized for metabolism. If utilization exceeds uptake, a state of O_2 debt exists, and body stores of O_2 are depleted. It is possible in this situation for the metabolic use of O_2, as measured by calorimetry, to be greater than the $\dot{V}O_2$ calculated by Fick or even measured by respiratory gas exchange.

Patients with the adult respiratory distress syndrome (ARDS) may represent another case in which O_2 uptake and utilization do not agree: they may not increase $\dot{V}O_2$ after increasing DO_2 (Ronco *et al.* 1991). Hanique *et al.* (1994) studied 3 groups of patients, septic, ARDS, and hepatic failure. They increased cardiac output by volume loading to increase DO_2, measured $\dot{V}O_2$ by calorimetry and calculated $\dot{V}O_2$ by Fick. They found that the calculated $\dot{V}O_2$ (Fick) increased, while the measured $\dot{V}O_2$ (calorimetry) did not. By using an increase in $\dot{V}O_2$ as a criterion for successful transfusion, they concluded that as many as 58% of transfusions may be of questionable importance (Babineau *et al.* 1992)

3.5.3 Oxygen Extraction Ratio (OER)

The oxygen extraction ratio is the fraction of arterial O_2 delivery extracted by tissue. In other words:

$$(4) \qquad OER = \frac{(CaO_2) - (C\bar{v}O_2)}{(CaO_2)}$$

This parameter has been useful in a number of animal studies. For example, Levy *et al.* (1992) created stenotic lesions in the left anterior descending coronary arteries of dogs and then studied their compensation to acute blood loss anemia. They found that the stenotic hearts did not raise their outputs in response to bleeding, had greater lactate production, and failed at a higher hematocrit (17%) than controls (10.6%). They found that in the normal heart, lactate production occurs when OER > 50% and hematocrit < 10%, but in the stenotic animals, OER > 50% corresponded to a hematocrit < 20%. These authors concluded that an OER > 50% indicates a need for transfusion, and stressed that the transfusion trigger, using any criteria, is higher in hearts with underlying coronary ischemia.

These observations are consistent with experience with patients. For example, Mathru *et al.* (1992) studied 8 patients with ejection fractions > 40% undergoing coronary artery bypass grafting procedures who were hemodiluted to a target hematocrit of 15%. Catheters were placed in the coronary sinus, and the hematocrit was increased by autologous transfusion, followed by measurements of hemodynamics and lactate production. They found that the OER decreased with increasing hematocrit, but lactate extraction was independent of hematocrit. There was no evidence of cardiac ischemia in any of the patients, and the authors concluded that the hematocrit was not as useful as a transfusion trigger as O_2 utilization or extraction.

3.6 Clinical Transfusion Triggers

Although these considerations provide some basis for a physiologic transfusion trigger, actual clinical situations are far more complicated. For example, not all patients can raise cardiac output in response to the challenge of anemia. In others, tissue ischemia, such as in coronary artery disease, can raise the local DO_2 requirement, reducing the O_2 reserve for that area. In other patients, pulmonary disease may impose a diffusion barrier (increased DMO_2) or restrict ventilation which can restrict pulmonary O_2 uptake.

Figure 3.5 shows some of the factors that can influence the O_2 delivery as described by the Fick equation. As already mentioned, these are highly iterative. It is beyond the scope of this discussion to explore each of them, but the examples and calculations above should make clear that the transfusion trigger will vary from patient to patient, is not quantifiable given our current state of quantitation, and the use of good clinical judgment cannot be avoided.

How then should the responsible physician consider the decision to transfuse? Table 3.2 presents an attempt at a rational approach. It is difficult to imagine a situation in which a hemoglobin over 10 g/dl would be desired. The issue of the optimal hematocrit has been explored in the literature extensively, and it appears that there is little justification for

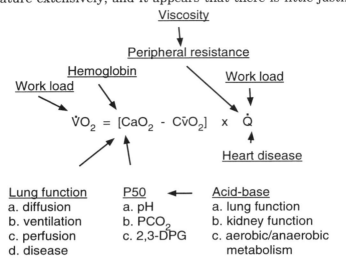

Figure 3.5. Interactions of the determinants of oxygen transport in health and disease (Winslow 1992).

Table 3.2 Clinical transfusion triggers.

Hemoglobin (g/dl)	Risk	Strategy
> 10	very low	avoid
8 - 10	low	avoid; transfuse if demonstrably better after trial
6 - 8	moderate	try to avoid; decrease $\dot{V}O_2$ Clinical evaluation: - volume status _ pulmonary status _ cardiac status (ischemia) - cerebrovascular disease - duration of anemia - dyspnea on exertion - estimated blood loss - extent of surgery, risk of rebleed
< 6	high	usually requires transfusion

Modified from Swisher and Petz (1989).

maintaining a hematocrit over 35% either at sea level or in high altitude natives (Winslow and Monge 1987). One reason for this is that as hematocrit rises, viscosity rises exponentially with increased resistance to flow and reduced cardiac output (Guyton, Jones and Coleman 1973).

When the hemoglobin is between 8 and 10 g/dl, the risk to most patients is very low. Some patients, especially elderly ones, report subjective improvement in symptoms of shortness of breath or dyspnea on exertion when their hematocrits are maintained over 8 g/dl. Transfusion in these patients would appear to be justified, but elevation to values over 10 g/dl would probably never be indicated.

A hemoglobin concentration between 6 and 8 g/dl requires a thoughtful approach to the clinical evaluation of the patient. One should try to avoid transfusion, and a number of alternatives are available such as lowering $\dot{V}O_2$ (*e.g.,* rest, pharmacologic agents, hypothermia) or treatments to modify the cause of anemia (*e.g.,* stop bleeding, treat underlying disease). But if neither of these can be done, then specific evaluation of a number of factors should be considered:

- Volume status. Is the patient hemodiluted or dehydrated? In other words, does the hemoglobin concentration reflect the true red cell mass?

- Pulmonary status. Is the patient able to oxygenate arterial blood? If not, why? Is the limitation diffusion or restricted ventilation? If the patient is a surgical candidate, will his ventilation be controlled?

- Cardiac status. Is there a history of myocardial ischemic disease or infarction? Such history would favor transfusion. Is the patient able to increase the cardiac output? The patient's age and symptoms are important factors here. The onset of coronary ischemic disease is most commonly 1-2 decades earlier in men than in women, and so sex also can be an important consideration. In general, one should assume coronary ischemia increases with age in both sexes.

- Cerebrovascular status. The considerations here are similar to those for cardiac status, except that the symptoms of ischemia may be more difficult to appreciate, particularly in the elderly. Is there a history of cerebrovascular accident? Are there neurological signs?

- Chronicity of anemia. An otherwise healthy person can adjust quite well to a hematocrit of 15% if the fall is slow, but if it is abrupt will usually cause severe symptoms.

- Symptoms. Does the patient complain of dyspnea on exertion, shortness of breath or claudication?

- Estimated blood loss. If a patient is undergoing a procedure in which the extent of blood loss is expected to be high, the O_2 reserve should be maximized.

- Extent of surgery, risk of rebleeding. In such cases (*e.g.*, coronary artery bypass graft procedures), rebleeding causes a significant risk of morbidity and mortality, especially in elderly, high-risk patients. Included here should also be any patients with increased risk for bleeding, such as thrombocytopenic or patients with liver disease.

When the hemoglobin is less than 6 g/dl, few would argue with the decision to transfuse *except* when the anemia is of very long standing. Such cases would include, for example, some patients with pernicious anemia who are well-adapted to a very low hematocrit. But the adaptation is due, in part, to chronically increased cardiac output and expanded blood volume, and too vigorous transfusion can push the patient into overt congestive heart failure.

3.7 Implications for Red Cell Substitute Development

If a red cell transfusion is given to *prevent* tissue hypoxia, how can a rational transfusion trigger be selected for clinical trials of red cell substitutes? When a transfusion is given to alleviate specific symptoms or signs of ischemia (ECG changes, shortness of breath, angina, *etc.*), then one might see evidence of improvement after transfusion and, conceivably, tests could be designed in which the end point would be to determine whether the indications are reversed.

Unfortunately, the optimal use of red cells is to prevent tissue hypoxia rather than to alleviate tissue ischemia, and so interpretation of clinical trials could be difficult if not impossible. It seems most likely that the best chance for a clear-cut demonstration of efficacy for red cell substitute will be in fully instrumented surgical patients whose detailed measurements of O_2 transport can be made. After efficacy is established in these patients, then the products could be used with more confidence in less intensively monitored patients.

3.8 Summary

A major goal of this discussion is to point out that physiological principles, while essential to understanding the rationale for transfusion, can provide only a general framework to consider the transfusion trigger. Within that framework, the decision to transfuse a given patient must be tailored to that patient's clinical condition. An understanding of both the parameters of O_2 transport in a patient and good clinical judgment will ultimately guide the thoughtful physician.

The design of clinical trials of blood substitutes should be applied to patients with clear-cut requirements for transfusion or to those in whom reasonable people can agree on a rational transfusion trigger. In the majority of patients in routine hospital practice, the decision to transfuse and the interpretation of data indicating the benefit of transfusion are much more complex. In these patients, a large amount of clinical data must be sorted, weighted, and sifted, and in the end the decision to transfuse is dominated by "good clinical judgment".

3.9 Acknowledgments

This work was supported by the National Heart, Lung and Blood Institute of the National Institutes of Health (P01 HL48018).

3.10 References

American College of Physicians. Practice strategies for elective red blood cell transfusion. *Ann. Int. Med.* 116: 403-406, 1992.

Babineau, T.J., W.H. Dzik, B.C. Borlase, J.K. Baxter, B.R. Bistrian, and P.N. Benotti. Reevaluation of current transfusion practices in patients in surgical intensive care units. *Am. J. Surg.* 164: 22-25, 1992.

Dodd, R.Y. The risk of transfusion-transmitted infection. *N. Engl. J. Med.* 327: 419-420, 1992.

Goodnough, L.T., D. Verbrugge, K. Vizmeg, and J. Riddell. Identifying elective orthopedic surgical patients transfused with amounts of blood in excess of need: the transfusion trigger revisited. *Transfusion* 32: 648-653, 1992.

Gould, S.A., A.L. Rosen, L.R. Sehgal, H.L. Seghal, L.A. Langdale, L.M. Krause, and G.S. Moss. O_2 extraction ratio: A physiologic indicator of transfusion need. *Transfusion* 23: 416(abstract), 1983.

Guyton, A.C., C.E. Jones, and T.G. Coleman. *Cardiac Output and Its Regulation. 2nd edition*. Philadelphia: Saunders, 1973.

Hanique, G., T. Dugernier, P.F. Laterre, A. Dougnac, J. Roeseler, and M.S. Reynaert. Significance of pathologic oxygen supply dependency in critically ill patients: comparison between measured and calculated methods. *Intensive Care Med.* 20: 12-18, 1994.

Levine, E., A. Rosen, L. Sehgal, S. Gould, H. Sehgal, and G. Moss. Physiologic effects of acute anemia: implications for a reduced transfusion trigger. *Transfusion* 30: 11-14 1990.

Levy, P.S., R.P. Chavez, G.J. Crystal, S.J. Kim, P.K. Eckel, L.R. Sehgal, H.L. Sehgal, M.R. Salem, and S.A. Gould. Oxygen extraction ratio: a valid indicator of transfusion need in coronary vascular reserve? *J. Trauma* 32: 769-773, 1992.

Lorente, J.A., L. Landin, R. DePablo, E. Renes, R. Rodriguez-Diaz, and D. Liste. Effects of blood transfusion on oxygen transport variables in severe sepsis. *Crit. Care Med.* 21: 1312-1318, 1993.

Lucking, S.E., T.M. Williams, F.C. Chaten, R.I. Metz, and J.J. Mickell. Dependence of oxygen consumption on oxygen delivery in children with hyperdynamic septic shock and low oxygen extraction [see comments]. *Crit. Care Med.* 18: 1316-1319, 1990.

Marik, P.E., and W.J. Sibbald. Effect of stored-blood transfusion on oxygen delivery in patients with sepsis. *J. A. M. A.* 269: 3024-3029, 1993.

Mathru, M., B. Kleinman, B. Blakeman, H. Sullivan, P. Kumar, and D.J. Dries. Myocardial metabolism and adaptation during extreme hemodilution in humans after coronary revascularization. *Crit. Care Med.* 20: 1420-1425, 1992.

Messmer, K., D.H. Lewis, L. Sunder-Plassmann, W.P. Klovekorn, N. Mendler, and K. Holper. Acute normovolemic hemodilution. Changes of central hemodynamics and microcirculatory flow in skeletal muscle. *Eur. Surg. Res.* 4: 55-70, 1972.

Mink, R.B., and M.M. Pollack. Effect of blood transfusion on oxygen consumption in pediatric septic shock [see comments]. *Crit. Care Med.* 18: 1087-1091, 1990.

Mollison, P.L.. *Blood Transfusion in Clinical Medicine.* 7th Edition. Philadelphia: FA Davis, 1983.

National Institutes of Health Consensus Conference. Perioperative Red Cell Transfusion. *J. Amer. Med. Assoc.* 260: 2700-2703, 1988.

Ronco, J.J., J.S. Montaner, J.C. Fenwick, and J.A. Russell. Pathologic dependence of oxygen consumption on oxygen delivery in a respiratory failure secondary to AIDS-related *Pneumocystis carinii* pneumonia. *Chest* 98: 1463-1466, 1990.

Ronco, J.J., P.T. Phang, K.R. Walley, B. Wiggs, J.C. Fenwick, and J.A. Russell. Oxygen consumption is independent of changes in oxygen delivery in adult respiratory distress syndrome [see comments]. *Am. Rev. Respir. Dis.* 143: 1267-1273, 1991.

Roughton, F.J.W., and R.E. Forster. Relative importance of diffusion and chemical reaction rates in determining rate of exchange of gases in the human with special reference to true diffusing capacity of pulmonary. *J. Appl. Physiol.* 11: 290-302, 1957.

Samaja, M., and R.M. Winslow. The separate effects of H^+ and 2,3-DPG on the oxygen equilibrium curve of human blood. *Br. J. Haem.* 41: 373-381, 1979.

Seear, M., D. Wensley, and A. MacNab. Oxygen consumption-oxygen delivery relationship in children. *J. Pediatr.* 123: 208-214, 1993.

Shoemaker, W.C., P.L. Appel, H.B. Kram, and T.-S. Lee. Comparison of two monitoring methods (central venous pressure versus pulmonary artery catheter) and two protocols as therapeutic goals (normal values versus values of survivors) in a prospective randomized clinical trial of critically ill surgical patients. *Crit. Care Med.* 13: 304, 1985.

Spiess, B.D., K.J. Tuman, R.J. McCarthy, W.G. Logas, T.R. Lubenow, J. Williams, H. Sankray, and P. Foster. Oxygen consumption and mixed venous oxygen saturation monitoring during orthotopic liver transplantation. *J. Clin. Monit.* 8: 7-11, 1992.

Stehling, L., and T.L. Simon. The red blood cell transfusion trigger. Physiology and clinical studies. *Arch. Pathol. Lab. Med.* 118: 429-344, 1994.

Swisher, S.N., and L.D. Petz. *Clinical Practice of Transfusion. 2nd Edition.* Churchill Livingstone, pp. 531-548, 1989.

Tomasulo, P.. World-wide blood supply and its impact on development of blood substitutes. In *Blood Substitutes: Physiological Basis of Efficacy* (R.M. Winslow, K. D. Vandegriff, and M. Intaglietta, eds.) Boston: Birkhäuser, 1995, pp. 1-19.

Wagner, P.D. Diffusion and chemical reaction in pulmonary gas exchange. *Physiol. Rev.* 57: 257-312, 1977.

Winslow, R.M.. A model for red cell O_2 uptake. *Int. J. Clin. Monit. Comput.* 2: 81-93, 1985.

Winslow, R.M. *Hemoglobin-based Red Cell Substitutes.* Baltimore: Johns Hopkins University Press, 1992.

Winslow, R.M., and C.C. Monge. *Hypoxia, Polycythemia, and Chronic Mountain Sickness.* Baltimore: Johns Hopkins University Press, 1987.

Winslow, R.M., M. Samaja, N.J. Winslow, L. Rossi-Bernardi, and R.I. Shrager. Simulation of the continuous O_2 equilibrium curve over the physiologic range of pH, 2,3-diphosphoglycerate, and pCO_2. *J. Appl. Physiol.* 54: 524-529, 1983.

Chapter 4

Combat Casualties, Blood, and Red Blood Cell Substitutes: A Military Perspective in 1995

Jon C. Bowersox, M.D., Ph.D. and John R. Hess, M.D., M.P.H.*

Vascular Surgery Service, Madigan Army Medical Center, Tacoma, Washington 98431-5000

*Blood Research Detachment, USAMRD-NNMC, Walter Reed Army Institute of Research, Washington, D.C. 20307

4.1 Introduction

Hemorrhage is the leading cause of death in soldiers injured on the battlefield (Bellamy 1984). Unlike other traumatic tissue loss, blood loss can be controlled and corrected. Effective first aid, resuscitation with crystalloid solutions and blood products, and surgical control of hemorrhage are the standard tools. Controlling hemorrhage will immediately reduce subsequent transfusion requirements and prevent the consequences of blood loss.

With severe blood loss or ongoing hemorrhage, maintaining the circulation and tissue perfusion requires replacement with crystalloid fluids, red blood cells, albumin, clotting factors, and platelets, in that order (Collins 1973). There is usually an abundant reserve in the body's oxygen delivery system in young people, thus volume replacement with oxygen-carrying solutions is generally not required in the initial phase of resuscitation.

Restoring intravascular volume with crystalloid or colloid solutions usually increases perfusion and restores oxygen delivery to peripheral tissues. If this is not effective, red blood cells are required. Red blood cells are recommended after the injured person has lost 30-40% of the circulating blood volume (1500-2000 ml in the average adult), although greater blood losses are tolerated under many circumstances (ATLS

Blood Substitutes: Physiological Basis of Efficacy
Winslow et al., Editors
© Birkhäuser Boston 1995

1989). Red blood cell transfusions are safe and highly effective in restoring oxygen-carrying capacity after severe hemorrhage. The long intravascular persistence of transfused red blood cells contributes to the hemodynamic stability of the injured during transport and recovery, and the viscosity and space-filling qualities of erythrocytes contribute to normal microvascular rheology and hemostasis.

Red blood cell substitutes are being developed to extend the availability of oxygen-carrying solutions to applications for which blood is not currently available and to replace blood in transfusion therapy (Winslow 1992). The oxygen-carrying capacity, universal infusion compatibility, and storage stability of these acellular oxygen carriers are seen as significant technical advances.

Potential applications for red blood cell substitutes exist in pre-hospital care, in certain hospital situations where time is critical, in mass casualty situations where the demand for red blood cells exceeds the immediate supply, and in elective surgical situations where the substitute can serve as a simple alternative oxygen carrier. However, to replace or augment blood use, red blood cell substitutes must compare favorably to the safety, efficacy, durability, and cost-effectiveness of crystalloid resuscitation solutions and blood. The comparison should reflect the conditions of anticipated use.

This paper briefly describes the epidemiology of military trauma, the historic patterns of blood use on the battlefield, the scientific indications for blood transfusion, and the U.S. Army's doctrine for blood use in the near future. We use this information to describe the logical constraints on the military usage of red blood cell substitutes.

4.2 The Epidemiology of Military Trauma

The incidence of injury on the battlefield is highly variable. The intensity of combat, the lethality of weapons, and the availability of cover all effect the incidence of wounds in measurable ways. In addition, there are the less easily measured qualities of military art and science. It is not uncommon to see very different casualty figures for opposite sides in the same battle. One has only to think of Agincourt, Fredericksburg, and the Persian Gulf for examples of the interplay of these factors.

Despite the large variation in the incidence of injury on individual battlefields, a clear pattern has emerged in the distribution in classes of casualties in major modern wars. Twenty percent of casualties were killed in action (KIA), dying on the battlefield before reaching a field hospital (Bellamy, Maningus and Vayer 1986). This proportion of casualties killed in action has remained unchanged for all the major wars Americans have fought since the Civil War. In contrast, the percentage of casualties who reach a hospital but then die of wounds (DOW) has decreased steadily from 28% in the Civil War to 2% in Vietnam (Bellamy, Maningus and Vayer 1986, Carey 1987, Garfield and Neugut 1991).

In practical terms, this means that most battlefield casualties are either killed outright or have survivable, frequently minor, wounds (Bellamy 1984). Those who are hospitalized generally require emergency surgery to control hemorrhage or to manage soft tissue, gastrointestinal, or brain injuries. In Vietnam, 45-92% of casualties were operated on shortly after admission (Byerly and Pendse 1981, McNamara and Stremple 1973, Arnold and Cutting 1978).

Reducing the time required to evacuate wounded combatants from the battlefield should reduce the percentage of casualties who die on the battlefield (Trunkey 1982). A result is that a greater proportion of fatally injured casualties will reach field hospitals alive. While overall hospital mortality has decreased with more rapid evacuation as a result of improved care, those who die usually do so shortly after arriving at the hospital. Hemorrhage and neurological injuries account for the overwhelming majority of immediate and overall hospital deaths.

Over 50% of casualties who are killed in action bleed to death (Bellamy 1984). Massive truncal hemorrhage accounts for most of these deaths. However, one analysis indicates that as many as 22% of fatal vascular injuries were located in the extremities, in areas in which the hemorrhage could have been controlled by the rapid application of first aid, *i.e.,* direct pressure (Bellamy 1984). In Vietnam, newsreel footage documents casualties with spurting arterial hemorrhage being loaded onto helicopters for evacuation to "definitive" care.

4.3 Blood Transfusion in Military Trauma

Blood transfusions and intravenous fluid infusions have been lifesaving in combat casualties since World War I (Robertson 1918). Blood is used to resuscitate casualties in severe shock, to replace blood perioperatively, and to correct subsequent anemia in casualties with deficits in red cell mass after fluid replacement with crystalloid solutions.

Most casualties who receive blood transfusions have hemorrhage requiring surgical control. Military doctrine has been developed to provide large quantities of blood to field hospitals, and a well organized logistics network (U.S. Armed Forces Blood Program) has been established to accomplish this task.

In Vietnam, 46% of all casualties admitted to field hospitals received blood transfusions (Mendelson 1975). Similar percentages were reported in the Falkland Islands campaign (64%) (Jowitt and Knight 1983) and during the civil war in Lebanon (50%) (Allam, Nassif and Alami 1983). In a recent report from the International Committee of the Red Cross (ICRC), only 16% of casualties treated in ICRC field hospitals in Thailand and Pakistan were transfused; however, almost 70% of the Cambodian and Afgani casualties described in this report arrived at hospitals more than six hours after being wounded (Eshaya-Chauvin and Coupland 1992).

The number of units of blood transfused per hospitalized casualty has been reported as 0.9-4.4 units during World War II, Korea, and Vietnam (Camp, Conte and Brewer 1973). This number is confounded by the inclusion of all hospitalized patients in the denominator, whether or not they were hospitalized for wounds. For planning purposes, a more useful figure is the number of units transfused per casualty receiving blood. In Vietnam, patients were transfused with an average of 4.3 units (Mendelson 1975). Similar values were reported from Lebanon (4.6 units/casualty) (Allam, Nassif and Alami 1983); however, only 2.9 units/casualty were administered by the ICRC field hospitals (Eshaya-Chauvin and Coupland 1992).

The amount of blood administered in field hospitals has depended on individual clinical practice, on clinical standards, and on the nature of wounds. Ringer's lactate used in Vietnam was called "Texas white blood" in reference to its almost exclusive use in resuscitation by a surgeon from Dallas (Shires and Canizaro 1973). The infrequency of single-unit transfusions reported in Vietnam - rates varied from 0 to 8% in several large series (Byerly and Pendse 1981, Mendelson 1975) - may reflect the use of blood in resuscitating severely hemorrhaging casualties before bleeding could be surgically controlled. The nature of injuries caused by land mines led to greater blood use in these casualties than in those caused by gunshot wounds or fragments from bombs or rockets.

Universal-donor (Type O, Rh positive) blood was widely used before 1966; most blood used subsequently has been fully cross-matched. The administration of universal-donor blood is extremely safe; more than 100,000 such units were given in Vietnam without a single fatal hemolytic transfusion reaction (Camp, Conte and Brewer 1973). These units represent at least 16% of all transfusions during that war. Typing and cross-matching can be performed within 20 minutes, but the inappropriate dependence on test-tube cross-matching in emergency situations in Vietnam led to patient identification errors, more than 40 hemolytic transfusion reactions, and nine recognized fatalities (Camp, Conte and Brewer 1973). Seventy percent of the 1.3 million units of blood sent to Vietnam were Type O, and more than half of all the blood sent to Vietnam outdated. Given the distribution of blood types in the American population and random outdating, this suggests that only 200,000 transfusions in non-Type O individuals could have been type-specific. Thus, the published rates probably underestimate the frequency of universal-donor transfusion in Vietnam and overestimate the safety of cross-matching in the typical circumstances of casualty care (Collins 1973).

4.4 The Indications for Transfusion

While the therapeutic objectives in bleeding patients are straightforward, the precise clinical indications that should trigger red blood cell transfusion in surgical care remain variable and controversial. They are further complicated by the quest for clear-cut medico-legal standards. The recent SANGUIS study of blood use in major teaching hospitals in Western Europe showed that use varied between hospitals from 0 to 100% for several common major surgical procedures (Baele 1994). Official doctrine in World War II and common practice in the Korean conflict defined procedure to transfuse patients to hematocrits around 40% to insure hemodynamic stability and wound healing (Kendrick 1964, Crosby 1988). Awareness of the infectious risk of blood and a better understanding of oxygen delivery have led to much more cautious use of blood in the 1990s.

In the United States, consensus conferences sponsored by the National Institutes of Health and position papers from professional organizations have discredited the old "transfusion trigger" of a hemoglobin below 10 g/dl or a hematocrit below 30% (NIH 1988, ACP 1992). New guidelines emerging from these conferences have not addressed trauma care specifically. NATO doctrine for military trauma transfusion follows guidelines established by the American College of Surgeons for the Advanced Trauma Life Support (ATLS) course (Bowen and Bellamy 1988, OTSG 1991). Blood transfusion is indicated for casualties with evidence of ongoing hemorrhage in the presence of shock and for those whose vital signs either fail to respond or respond only transiently to volume infusion. These guidelines also recommend administering blood to casualties bleeding more than 100 ml/min.

Several authors recommend hemoglobin concentrations below 7 g/dl (OTSG 1991) or 8 g/dl (Allam, Nassif and Alami 1983) as "triggers" for transfusion. However, criteria based on hemoglobin concentration suffer from the fact that the minimally acceptable hemoglobin concentration is an individual characteristic that depends on non-hemoglobin variables, including the ability to increase cardiac output, tissue oxygen demand, pH, the ability to oxygenate available hemoglobin, and the adequacy of perfusion to critical vascular beds (Lundsgaard-Hansen 1992). Clinical studies show that some healthy young people tolerate general anesthesia and surgery with hemoglobin concentrations close to the critical oxygen delivery threshold of 4 g/dl (Spense 1991) while older patients with vascular disease can show electrocardiographic abnormalities at hemoglobin concentrations of 10 g/dl, which improve with red blood cell transfusion (Czer and Shoemaker 1978).

Another group of proposed transfusion criteria are based on the mixed venous oxygen saturation, such as a mixed venous oxygen saturation less than 67% (Lundsgaard-Hansen, Doran and Blauhut 1989) or an oxygen extraction ratio of greater than 50% (Wilkerson *et al.* 1987). These values, which correspond under normal circumstances to pulmonary artery oxygen tensions of 35 and 30 torr, have very limited ability to discriminate whether or not additional red blood cells will improve oxygen consumption at hemoglobin concentrations above 7 g/dl. As blood gas determinations are not available in most field hospital situations, the decision to transfuse emergently remains a clinical judgement based on physical findings and the hemoglobin.

Hemoglobin concentration does not have the same meaning in acutely hemorrhaging trauma patients that it has in more stable surgical patients. High hemoglobin concentrations or hematocrits in combat casualties often underestimate the acute intravascular volume loss, as demonstrated in the Danang Naval blood utilization study where the mean admission hematocrit was 36.5 ± 5.3% (McCaughey *et al.* 1988). Low hemoglobin concentrations that are well tolerated in study models, such as anemic volunteers or animals who are otherwise healthy, may not be safe for military trauma casualties. Additional encroachments on tissue oxygen supply may result from increased cellular metabolic requirements, arterial hypoxemia, or alkalosis, leaving little physiologic reserve at lower hemoglobin concentrations. The availability of supportive measures that could increase tissue oxygen delivery, such as oxygen and ventilators, is likely to be limited in the austere settings of field hospitals.

The decision to transfuse trauma patients is a clinical decision embedded in the emergency context in which the usual objective laboratory measures can mislead. A recent review of blood use in 1992 in all Class I trauma centers in Illinois showed that 18% of patients who received any blood transfusion received only a single unit of red blood cells, and that 53% received two units or less (Gould *et al.* 1994). Thus, in these civilian centers where 85% of the injuries are blunt trauma, the decision to start transfusing is shortly followed by the decision to stop. The current ATLS blood transfusion decision algorithm used in these centers may lead to some unnecessary blood transfusion, but blood is a powerful tool to correct the signs of hemorrhagic shock, and the clinical price of untreated shock is high.

4.5 Planning for Military Blood Use

At the height of the Cold War, plans for military blood use were based on predictions of high-intensity combat involving field armies with millions of soldiers on the plains of northern Europe. Blood requirements of more than 100,000 units a day were estimated. At that rate, blood use would have been more than three times the sustained yield of the entire U.S. blood banking system (Wallace, Surgenor and Hao 1993). Additional logistic constraints required that blood be available on short notice and that it not monopolize limited airlift capabilities. These issues became the driving forces in the development of alternatives to liquid blood. As the Soviet military threat has diminished, U.S. military blood requirements have been markedly reduced. Requirements to deploy blood with U.S. troops around the world remain.

Estimates of blood use based on previous U.S. military experience suggest far more modest blood requirements. Even the largest demands for blood faced in 1968, when 476,000 units were shipped to Vietnam, or in 1991, when 120,000 units were shipped to the Persian Gulf, represent less than 4% and 1%, respectively, of the annual blood supply in the United States. Although 60% of the blood shipped to Vietnam and 85% of the blood sent to the Persian Gulf became outdated, the costs to provide guaranteed availability are willingly borne by military planners. The total number of red blood cell units that have outdated to meet U.S. military contingency requirements over the last 50 years appears to be approximately 1 million units.

Advances in medical knowledge and technology can be expected to further reduce military requirements for blood. Knowledge of patterns of combat death has shaped training doctrine for soldiers and medics to emphasize of the importance of good first aid for hemorrhage control to reduce primary blood loss. Resuscitation research has pointed out the importance of avoiding excessive fluid replacement in hemorrhaging patients (Bickell *et al.* 1994).

Reducing fluid administered will limit blood loss and blood dilution. Trauma surgeons have developed a set of techniques collectively called "damage control surgery" with emphasis on rapid hemorrhage control and avoidance of the coagulopathy of hypothermia (Rotondo *et al.* 1993). Tissue adhesives, such as the fibrin glue and fibrin foam used successfully for hemorrhage control in World War II (Kendrick 1964), are being redeveloped now that the infectious disease risks of such products appear preventable.

Computer management of blood on the battlefield and computer crossmatching of blood for transfusion on the battlefield both will increase the efficiency of blood use. Lastly, improvements in liquid blood storage will reduce blood outdating.

4.6 Red Blood Cell Substitutes in Military Trauma

The efficacy of red blood cell transfusions in the treatment of hemorrhagic shock has been well substantiated for over half a century. The safety of both universal-donor and type-specific blood and the recovery and survival of red blood cells in banked blood is likewise well documented.

Military surgical management and blood requirements are clearer and more readily met than at any time in history. In this context, a red blood cell substitute must have good oxygen-carrying capacity, universal-recipient compatibility, and long shelf storage and be limited only by manageable side effects and short intravascular persistence. But even so, it cannot replace red blood cells on the battlefield, because it may not fulfill all of the functions of transfused red blood cells in combat casualty care. There is no evidence for instance that in a severely injured and massively transfused casualty, any such red blood cell substitute can provide hemodynamic stability and hemostasis. But if red blood cells are still needed on the battlefield, then much of the logistic benefit of fielding a substitute is lost. And if more than one unit of the substitute is required to replace each unit of blood because of short intravascular persistence, then the benefit diminishes further.

In the pre-hospital setting, oxygen-carrying solutions must be better than crystalloid solutions at prolonging life in a situation where oxygen-carrying capacity is almost always adequate. Control of hemorrhage becomes the critical action in this situation. The benefits of additional oxygen transport will be marginal at best and difficult to prove.

The ability to provide safe blood on the battlefield is better now than it has ever been. The total cost in outdated units of providing this visible support to America's soldiers has been about 1 million units over the last fifty years, or about 20,000 units a year. This is a small amount in a country that uses 12 million units of blood a year. The extent to which a safe red blood cell substitute might avoid this waste of blood and money remains entirely theoretical at present.

The U.S. Army has contributed extensively to the development of red blood cell substitutes and has benefited greatly from the better understanding of blood function that has resulted. Producing a useful red blood cell substitute remains an Army technical objective. Meanwhile, the care of American soldiers wounded in combat can be accomplished with crystalloid fluids and human blood products, and the orderly development and testing of red blood cell substitutes can continue.

4.7 Acknowledgements

The opinions and assertions contained herein are the private views of the authors and are not to be construed as official nor do they reflect the views of the Department of the Army or the Department of Defense (AR360-5).

4.8 References

(ACP) **American College of Physicians.** Practice strategies for elective red blood cell transfusion. *Ann. Intern. Med.* 116: 393-402, 1992.

(ATLS) Advanced Trauma Life Support Course for Physicians. Chicago: American College of Surgeons, 1989, pp. 59-73.

Allam, C.K., R.E. Nassif, and S.Y. Alami. Disaster transfusion experience. *Mid. East. J. Anesth.* 7: 147-152, 1983.

Arnold, K., and R.T. Cutting. Causes of death in United States military personnel hospitalized in Vietnam. *Milit. Med.* 143: 161-164, 1978.

Baele, P. Transfusion depends on the doctor, not on the patient: the SANGUIS study of transfusion in elective surgery in Europe [editorial]. *Acta Anaesthesiol. Belg.* 45: 3-4, 1994.

Bellamy, R.F. The causes of death in conventional land warfare: implications for combat casualty care research. *Milit. Med.* 149: 55-62, 1984.

Bellamy, R.F., P.A. Maningas, J.S. Vayer. Epidemiology of trauma: military experience. *Ann. Emer. Med.* 15: 1384-1388, 1986.

Bickell, W.H., M.J. Wall, P.E. Pepe, R.R. Martin, V.F. Ginger, M.K. Allen, and K.L. Mattox. Immediate versus delayed resuscitation for hypotensive patients with penetrating torso injuries. *N. Engl. J. Med.* 331: 1105-1109, 1994.

Bowen, T.E., and R.F. Bellamy (eds.). *Emergency war surgery.* Washington, D.C.: U.S. Government Printing Office, 1988, pp. 134-148.

Byerly, W.G., and P.D. Pendse. War surgery in forward surgical hospitals in Vietnam: a continuing report. *Milit. Med.* 143: 161-164, 1981.

Camp, F.R., N.F. Conte, and J.R. Brewer. *Military blood banking 19411973.* Fort Knox, KY: U.S. Army Medical Research Laboratory, 1973, p. 20.

Carey, M.E. Learning from traditional combat mortality and morbidity data used in the evaluation of combat medical care. *Milit. Med.* 152: 6-13, 1987.

Collins, J.A. Massive transfusion and current blood banking practices. In *Preservation of Red Blood Cells.* (H. Chaplin, E.R. Jaffe, C. Lenfant, and C.R. Valeri, eds.). Washington D.C.: National Academy of Sciences, 1973, pp. 39-40.

Crosby, W.H. Acute anemia in the severly wounded battle casualty. *Milit. Med.* 153: 25-27, 1988.

Czer, L.C.R., and W.C. Shoemaker. Optimal hematocrit value in critically ill postoperative patients. *Surg. Gynecol. Obstet.* 147: 363-368, 1978.

Eshaya-Chauvin, B., and R.M. Coupland. Transfusion requirements for the management of war injured: the experience of the International Committee of the Red Cross. *Br. J. Anaesth.* 68: 221-233, 1992.

Garfield, R.M., and A.I. Neugut. Epidemiologic analysis of warfare. *J. Am. Med. Assoc.* 266: 688-692, 1991.

Gould, S.A., L.R. Sehgal, H.R. Sehgal, and G.S. Moss. The role of hemoglobin solutions in massive transfusion. In *Massive Transfusion.* (L.C. Jefferies and M.E. Brecher, eds.) Bethesda MD: American Association of Blood Banks, 1994, pp. 43-64.

Jowitt, M.D., and R.J. Knight. Anesthesia in the Falklands campaign. *Anaesth.* 38: 776-783, 1983.

Kendrick, D.B. *Blood program in World War II.* Washington, D.C.: Office of the Surgeon General, 1964.

Lundsgaard-Hansen, P. Treatment of acute blood loss. *Vox Sang.* 63: 241-246, 1992.

Lundsgaard-Hansen, P., J.E. Doran, and B. Blauhut. Is there a generally valid minimally acceptable hemoglobin level? *Infusiontherapie* 16: 167-175, 1989.

McCaughey, B.G., J. Garrick, L.C. Carey, and J.B. Kelley. Naval support activity hospital, Danang, casualty blood utilization, January to June 1968. *Milit Med.* 153: 181-185, 1988.

McNamara, J.J., and J.F. Stremple. Causes of death following combat injury in an evacuation hospital in Vietnam. *J. Trauma* 12: 1010-1012, 1973.

Mendelson, J.A.. The use of whole blood and blood volume expanders in U.S. military medical facilities in Vietnam, 1966-1971. *J. Trauma* 15: 1-13, 1975.

(NIH) National Institutes of Health Consensus Development Conference. Perioperative red blood cell transfusion. *J. Amer. Med. Assoc.* 260: 2700-2703, 1988.

(OTSG) Combat Casualty Care Guidelines. *Operation Desert Storm.* Washington, D.C.: Office of the Surgeon General, 1991, pp. 30-33.

Robertson, O.H. Transfusion with preserved red blood cells. *Brit. Med. J.* 1: 691-695, 1918.

Rotondo, M.F., C.W. Schwab, M.D. Mcgonigal, G.R. Phillips, T.M. Fructerman, D.R. Kauder, B.A. Latenser, and P.A. Angood. 'Damage control': an approach for improved survival in exsanguinating penetrating abdominal injury. *J. Trauma* 35: 375-383, 1993.

Shires, G.T., and P.C. Canizaro. Fluid resuscitation in the severely injured. *Surg. Clin. North Am.* 53: 1341-1366, 1973.

Spence, R.K. The status of bloodless surgery. *Transfus. Med. Rev.* 5: 274-286, 1991.

Trunkey, D.D. Overview of trauma. *Surg. Clin. North Am.* 62: 3-7, 1982.

Wallace, E.L., D.M. Surgenor, H.S. Hao, J. An, R.H. Chapman, and W.H. Churchill. Collection and transfusion of blood and blood components in the United States, 1989. *Transfusion* 33: 139-144, 1993.

Wilkerson, D.K., A.L. Rosen, S.A. Gould, L.R. Segal, H.L. Segal, and G.S. Moss. Oxygen extraction ratio: a valid indicator of myocardial metabolism in anemia. *J. Surg. Res.* 42: 629-634, 1987.

Winslow, R.M. *Hemoglobin-based red blood cell substitutes.* Baltimore: Johns Hopkins University Press, 1992.

Chapter 5

Clinical Development of Perfluorocarbon-based Emulsions as Red Cell Substitutes

Robert J. Kaufman, Ph.D.

HemaGen/PFC, 11810 Borman Drive, St. Louis, Missouri 63146

ABSTRACT

The use of PFCs in medicine has been the subject of much research over the past 30 years. Recent technological progress in second generation oxygen transport products has solved many of the problems of Fluosol DA® including PFC concentration, elimination of surfactant side effects and improved storage stability. Some side effects, most notably, thrombocytopenia and flu-like symptoms still remain to be solved. Two new oxygen transport formulations, Oxyfluor™ and Oxygent™ are in clinical trials at this time. Several applications of PFCs have been approved for use by the FDA but have languished in the market place. Prominent among these is the application of Fluosol DA® to PTCA and Imagent GI for MRI contrast in the bowel. Other applications such as cancer therapy, ultrasound contrast and direct ^{19}F MR imaging continue to show promise.

5.1 Introduction

The search for solutions to temporarily replace the oxygen transport function of blood has been underway in academia, the military and industry for close to 50 years. Recent advances in perfluorocarbon emulsion technology indicate that the problems which have impeded development of useful medical products in this field may now be moving towards solution.

The historical background, mechanism of oxygen transport and fundamental aspects of perfluorocarbons and their emulsions has recently been reviewed (Riess and Leblanc 1978, Riess 1984, Kaufman 1991,

Blood Substitutes: Physiological Basis of Efficacy
Winslow et al., Editors
© Birkhäuser Boston 1995

Kaufman 1992). This review will focus on new advances in the clinical development of perfluorocarbon emulsions.

5.2 Applications

5.2.1 Temporary Oxygen Transport

Since the pioneering experiments of Clark, Sloviter and Geyer (Clark and Gollan 1966, Sloviter and Kamimoto 1967, Geyer, Monroe and Taylor 1968), much research has been devoted to developing medically useful oxygen transport products based on perfluorocarbons (PFCs) to temporarily replace the oxygen transport function of whole blood. Green Cross was the first company to reach clinical trials with an emulsion containing two perfluorocarbons, 70% perfluorodecalin and 30% perfluorotripropylamine, using pluronic as the primary surfactant. This emulsion, called Fluosol DA®, contained 10% PFC by volume and had to be stored in the frozen state (Riess 1984).

Clinical results with Fluosol DA® in oxygen transport were discouraging. In studies of severely anemic surgical patients (Hb <10) who refused transfusions on religious grounds, Fluosol failed to provide more than a transient boost in arterial oxygen tension (Gould et al. 1986). Fluosol DA® did contribute 28% of the oxygen consumption of these patients. However, this gain in oxygen consumption due to the PFC was offset by dilution of the hemoglobin. After infusion of Fluosol DA®, the net oxygen carried by the red cells, the plasma and the PFC was the same as that carried by the red cells and plasma before infusion (Table 5.1).

Table 5.1 Hemodynamics and oxygen transport properties before and after Fluosol DA® administration in eight patients[1,2].

Property	Before Fluosol DA®	After Fluosol DA®
Oxygen Delivery (mL/min/m^2)	235±27	197±32
Oxygen Consumption (mL/min/m^2)	109±13	88±11
Arterial Oxygen Tension (Torr)	356±24	430±19[3]
Venous Oxygen Tension (Torr)	40±3.9	78.2±3

[1]Values given as the mean±SE.
[2]Data were obtained at peak arterial oxygen content after Fluosol DA® administration.
[3]The difference between values before and after Fluosol DA® is significant (p<0.05).
Data from Gould et al. (1986).

The conclusions from Gould's study were that Fluosol DA® was safe but ineffective. In order to achieve a circulating PFC concentration of 5% required an infusion of 2800 mL of Fluosol DA®, 90% of which is aqueous. Therefore, concentrated emulsions would be necessary to achieve a PFC concentration at which significant oxygen transport occurs without diluting hemoglobin.

Additional clinical trials revealed that complement activation occurred in 5-10% of the patients treated with Fluosol DA®. The surfactant, pluronic F-68, was subsequently implicated as the cause of this effect (Tremper, Vercellotti and Hammerschmidt 1984). The clinical experience with the first generation product, Fluosol DA®, demonstrated that future products needed to be more concentrated, free from pluronic surfactants and possess better storage stability.

Two new products in development appear to have solved these problems: Oxyfluor™ from HemaGen/PFC and Oxygent™ from Alliance Pharmaceutical.

Oxyfluor™ is a 40% v/v emulsion of perfluorodichlorooctane (PFDCO) stabilized using egg yolk phospholipid and safflower oil. In equilibrium with 100% oxygen at 37°C, Oxyfluor™ carries 17.2 volume % oxygen. Oxyfluor™ has a shelf-life of greater than one year at room temperature.

PFDCO has a molecular weight of 471, boils at 155°C and has a density of 1.76 g/mL. It is a lipophilic PFC and is soluble in many hydrocarbons and other organic solvents. The ozone depletion potential (ODP) of PF-DCO is < 0.25.

PFDCO was discovered in a collaboration between HemaGen and 3M scientists when animal studies using perfluorodecalin (PFD) and perfluorooctyl bromide (PFOB) demonstrated that some PFCs caused pulmonary hyperinflation in rats and dogs. The phenomenon is characterized by an increase in respiratory rate, an increase in functional residual capacity, a decrease in inspiratory capacity and a reduction in arterial oxygen tension. Gross necropsy in rats or dogs at the time of peak effect showed lungs which remained inflated after removal from the chest. In rats, the lung volume from PFD treated rats measured by water displacement is two to three times control. Histologic evaluation of hyperinflated lungs did not reveal any structural defects. Except in extreme cases such as perfluorodimethlycyclohexane, this effect is not lethal in rats. Pulmonary hyperinflation was completely reversible with time, generally peaked at four to seven days post-infusion and disappeared completely by 30 - 60 days post-infusion.

A key problem in identification of new candidate molecules was to find PFCs that did not cause pulmonary hyperinflation but retained sufficient vapor pressure to leave the body in a reasonable time frame. To identify compounds with both properties, new PFCs were tested in rats for lung inflation by measurement of lung volume by water displacement and for liver clearance by gas chromatographic analysis of tissue at 2 and 14 days post-infusion. The results from a comparison of the lung volumes induced by PFD, PFOB, PFDCO and saline in rats is shown in Table 5.2.

Table 5.2 Effect of three PFCs on lung inflation in the rat.

Treatment	Gross Appearance	Lung Volume[1]
PFD	Marked Hyperinflation	2.20
PFOB	Modest Hyperinflation	1.10
PFDCO	Normal, Collapsed	0.90
Saline	Normal, Collapsed	0.87-0.91

[1]mL/100 grams of body weight. Values are mean of ten rats.

The results of the rodent tissue clearance studies for PFDCO are shown in Figure 5.1. The liver half-life is about 8 days and the PFDCO drops below the level of detection by 60 days post-infusion.

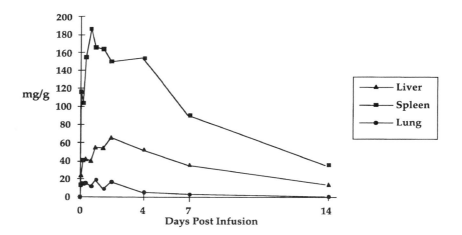

Figure 5.1 Clearance rate of PFDCO from the liver, spleen and lungs.

Candidate PFCs that passed the rodent tests were then tested in baboons for effects on pulmonary function as well as vital signs, hematology, coagulation and clinical chemistry. Historically, increases in respiratory rate correlate well with a decrease in inspiratory capacity and arterial blood gas tension, and an increase in functional residual capacity. PFDCO was compared to PFOB and PFD since they were the most widely studied PFCs in the field. PFD, which exhibited the largest rodent lung volume, had increased respiratory rate to 200% of baseline by day one post-infusion in the baboon. PFOB, which marginally inflated lungs in rats, surprisingly increased respiratory rate to 400% of baseline by four days post-infusion in the baboon and still exhibited elevated respiratory rate 60 days post-infusion (Figure 5.2). Respiratory rate in PFDCO treated baboons rose only to 140% of baseline by seven days post infusion and was in the normal range thereafter. PFDCO had only modest (≤20%) changes in inspiratory capacity, functional residual capacity and arterial blood gases. PFD and PFOB were similar in their marked effect on pulmonary function (Kaufman 1992). Baboons treated with these PFCs had a 35% decrease in inspiratory capacity and a 200% increase in functional residual capacity between days four and seven post infusion. Blood gas oxygen tension fell below 60 mm Hg at day four post-infusion in animals treated with PFD or PFOB, while PFDCO treated animals remained in the normal range throughout the post-infusion period.

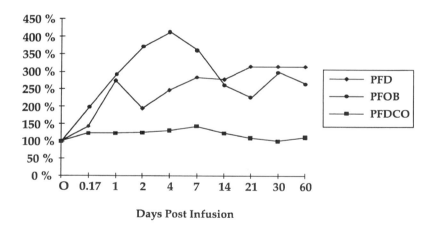

Figure 5.2 Comparison of PFD, PFOB and PFDCO on respiratory rate as a percent of baseline in the baboon at 8 ccPFC/kg.

Pre-clinical animal studies with Oxyfluor™ in models of shock resuscitation (Goodin *et al.* 1994) and surgical anemia (Kaufman 1992) have shown efficacy for oxygen transport. In contrast to traditional shock models, this model utilized compromised tissue oxygenation as an endpoint. Groups of dogs were hemorrhaged to a mixed venous oxygen tension of ≤25 mm Hg and held there for 10 minutes. They were resuscitated with either Oxyfluor™ (4 ccPFC/kg or 15 cc/kg of emulsion) or lactated Ringer's solution (15 mL/kg) while breathing 100% oxygen. Dogs resuscitated with Oxyfluor™ had normal mixed venous oxygen tension post-resuscitation and all survived. Dogs resuscitated with lactated Ringer's solution only returned 80% of a normal mixed venous oxygen tension and only 62.5% survived (Figure 5.3).

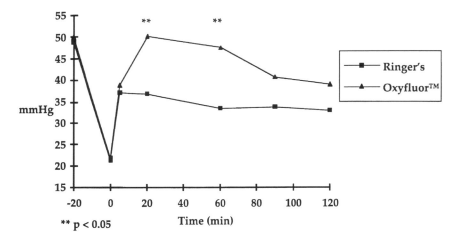

Figure 5.3 Effect of 4 ccPFC/kg of Oxyfluor™ on mixed venous oxygen tension in shocked dogs.

In the model of surgical anemia, dogs were hemodiluted to a hematocrit of 20% and then infused with Oxyfluor™ and allowed to breath 100% oxygen. Following the administration of Oxyfluor™, the arterial oxygen tension, venous oxygen tension and oxygen consumption were all increased significantly for 90 minutes. Results show that 40% of the consumed oxygen was derived from the circulating PFC for up to 90 minutes post-infusion. This dramatically illustrates the ability of PFCs to deliver significant amounts of oxygen to tissues (Figure 5.4).

A Phase I clinical trial in healthy human volunteers has been completed with Oxyfluor™. This trial involved three dosing levels with six treatment and six control subjects at each dosage level. The study was randomized within each dosing level and was double blinded. The subjects were monitored for vital signs, hemodynamics, clinical chemistry, hematology, coagulation, pulmonary function and pharmacokinetics for 30 days post-infusion. There were no reportable adverse events, bleeding abnormalities, respiratory abnormalities, cardiac abnormalities or complement activation.

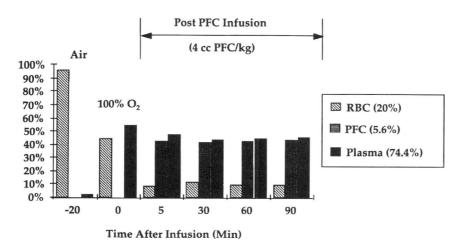

Figure 5.4 Fractional contribution to oxygen consumption of 4 ccPFC/kg of Oxyfluor™ in anemic dogs.

Dose related flu-like symptoms were observed four hours post-infusion and included fever, chills, nausea, leukocytosis and PMN shift, increased heart rate and lowered diastolic blood pressure. Only the high dose group showed consistent fever, which peaked at 101°F eight hours post-infusion (Figure 5.5).

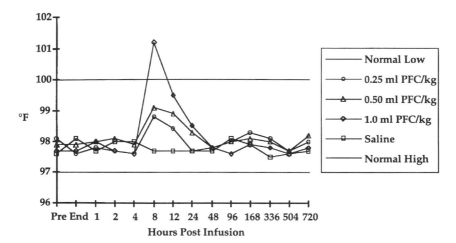

Figure 5.5 Effect of Oxyfluor™ on body temperature (°F) of human subjects after infusion at various doses.

The diastolic blood pressure dropped in the high dose group to a nadir of 55 mm Hg at four to 12 hours post-infusion (Figure 5.6), while the high-dose group heart rate rose to 105 beats per minute in the same time frame (Figure 5.7). The flu-like symptoms resolved by 24 hours post-infusion without intervention and appeared to be the direct consequence of phagocytosis of the PFC particles.

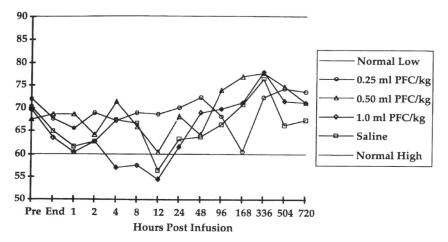

Figure 5.6 Effect of Oxyfluor^TM on disastolic pressure (mm Hg of human subjects after infusion at various doses.

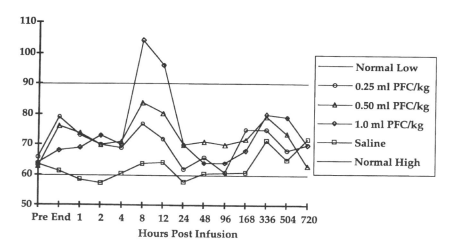

Figure 5.7 Effect of Oxyfluor^TM on heart rate (beats/minute) of human subjects after infusion at various doses.

The leukocytosis that accompanied the flu-like symptoms began at four hours in the high dose group and peaked at 24 hours post infusion (Figure 5.8).

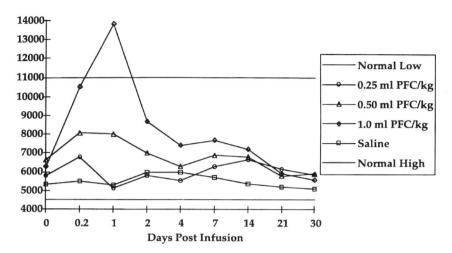

Figure 5.8 Effect of Oxyfluor^{TM} on white cells (WBC/µL) of human subjects after infusion at various doses.

In addition, a dose responsive, mild thrombocytopenia was observed two days post-infusion (Figure 5.9) with a nadir of 125,000/mL. Platelet numbers returned to normal by four days post-infusion. There was no evidence of bleeding associated with the drop in platelet count.

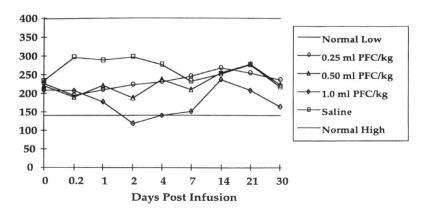

Figure 5.9 Effect of three doses of Oxyfluor^{TM} on platelet number (X1000) over time.

The pharmacokinetics of Oxyfluor™ were dose responsive with a blood half-life of two hours at 1 ccPFC/kg as measured by gas chromatographic analysis of subjects' blood.

Animal studies have indicated that the flu-like symptoms triggered by Oxyfluor™ (and other PFC emulsions) appear to be due to phagocytosis of the particles by macrophages and subsequent release of cytokines and arachadonic acid metabolites. Pre-treatment of mice with dexamethasone (Kaufman 1994) followed by 8 ccPFC/kg of Oxyfluor™ resulted in a significant reduction in the quantity of circulating IL-6 (Figure 5.10).

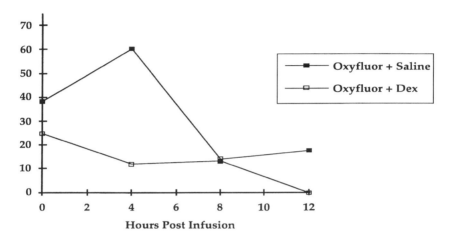

Figure 5.10 Effect of 0.1 mg/kg of dexamethasone on IL-6 (ng/mL) production induced by Oxyfluor™ in mice.

Subsequent clinical investigations will focus on the safety of Oxyfluor™ in surgical patients.

The other PFC based product in clinical development, Oxygent™ is a 90 w/v% (46 v/v%) emulsion of PFOB using egg yolk phospholipid. The emulsion is reported to be stable at room temperature for at least one year. PFOB is lipophilic, has a molecular weight of 499, boils at 141°C and has a density of 1.92 g/mL. The ODP of PFOB is between 0.25 and 2. PFOB has tissue clearance characteristics similar to PFDCO. Oxygent™ has been tested in dog models of surgical anemia with similar results to Oxyfluor™ (Cernaianu et al. 1993). Clinical trials of Oxygent™ have been completed in both healthy human volunteers and surgical patients (Keipert 1994, Spence 1994). Flu-like symptoms were also observed at 4-6 hours post-infusion, which were resolved by 24 hours post-infusion. Thrombocytopenia was also observed in these patients at the highest dose with the nadir of 140,000/mL occurring two days post-infusion.

A study of a similar PFOB-based product, Imagent BP, demonstrated reversal of flu-like symptoms in swine using dexamethasone, ibuprofen or indomethacin (Flaim *et al.* 1991).

5.2.2 Elimination of Gaseous Microemboli during Cardiopulmonary Bypass (CPB) Surgery.

For years there have been anecdotal reports of neurological and neuropsychological deficits following CPB. In a landmark study, Shaw, using a battery of neurological and neuropsychological tests quantitatively defined the extent and timing of the problem (Shaw *et al.* 1987). The results indicated that seven days post-CPB, 61% of patients have neurological deficits and 79% have neuropsychological deficits and that one year later, the deficits had persisted in about half of these patients (17% and 38%, respectively). The cause of these deficits have not been documented but are believed to be either platelet microaggregates caused by the circulation through the CPB apparatus or gaseous microemboli (GME) generated during cannulation, hypothermia and oxygenation of the blood. In 1990, two groups developed data that strongly suggested that a significant portion of these deficits were due to GME.

Taylor used retinal fluorescein angiography to compare patients who had bypass with bubble oxygenators to patients who had bypass with membrane oxygenators (Blauth *et al.* 1990). There was more than a 50% reduction in lesions in the membrane-oxygenator patient group. Moody performed autopsies on patients recently expired post-CPB and found numerous small capillary arteriole dilations in the brain which were thought to be due to GME (Moody *et al.* 1990). In contrast, brains from patients having non-CPB procedures did not have these lesions.

Spiess found that PFC emulsions could protect animals from cerebral emboli in a decompression model in rats (Spiess *et al.* 1988). This work suggested that PFC emulsions might be useful in eliminating GME from CPB patients and result in an improvement in their neurological and neuropsychological outcome. Using retinal angiography, Taylor found that Oxyfluor™ significantly reduced the frequency of GME and the area of circulatory dropout in dogs undergoing CPB (Taylor *et al.* 1992). The data, shown in Figure 5.11, clearly illustrates that low doses of Oxyfluor™ (1 ccPFC/kg) produce a ten-fold reduction in GME.

5.2.3 Percutaneous Transluminal Coronary Angioplasty (PTCA).

PFCs have been found to eliminate the transient myocardial ischemia induced during balloon inflation during PTCA in animal models (Spears *et al.* 1983) and in human clinical trials (Cleman, Jaffe and Wholgelernter 1986, Jaffe *et al.* 1988). In the clinical trial, symptomatic patients with single lesions of 70% or greater stenosis of the coronary artery undergoing PTCA were studied. All patients underwent a preliminary inflation without perfusion. Subsequent inflations were done with either oxygenated lactated Ringer's, oxygenated Fluosol DA® or unoxygenated Fluosol DA®. The data showed that oxygenated Fluosol DA® maintained an ejection fraction identical with the baseline 45 seconds post-balloon inflation, while there was a 35% reduction in ejection fraction in the controls. Fluosol DA® was approved by the US FDA in December in 1989 for use as an adjunct in high risk PTCA patients. Unfortunately, improved catheter technology has reduced the market for this product to such an extent that the manufacturer has withdrawn the product from the market.

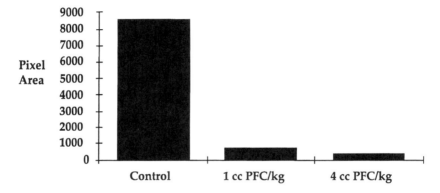

Figure 5.11 Effect of 1 and 4 ccPFC/kg of Oxyfluor™ on capillary dropout in dogs undergoing CPB.

5.2.4 Myocardial Infarction (MI)

Numerous studies have suggested that PFCs have beneficial effects in animal models of MI. Generally these studies have investigated two different approaches to treating MI: as a method of oxygenating the myocardium distal to the coronary artery occlusion (Glogar *et al.* 1981, Nunn *et al.* 1983) or as an adjunct to thrombolysis to prevent reperfusion injury (Bajaj *et al.* 1989). The dog studies of reperfusion injury showed a 60% reduction in the ratio of the area of necrosis to the area at risk (A_N/A_R). Mechanistically, this was attributed to a reduction in circulating neutrophils and reduced neutrophil chemotactic ability in the Fluosol DA®-treated dogs. It is still not known if this is just an effect of the pluronic surfactant.

A large clinical trial of Fluosol DA® in reduction of reperfusion injury, TAMI - 9, has been completed (Wall *et al.* 1994). In this study, 430 patients with acute MI symptoms of less than 6 hours duration and ST elevations of at least 0.1 mV in two of six leads were randomized into two groups. The first group received 100 mg tissue plasminogen activator (TPA) and 15 mL/kg of Fluosol DA®, while the second group received TPA alone. The clinical endpoints were global ejection fraction, regional wall motion, infarct size and clinical outcome. There were no significant differences in global ejection fraction, wall motion or infarct size between the two groups. There was a significant increase in pulmonary edema (45 vs. 31%) and a reduction in the incidence of recurrent ischemia (6 vs. 11%) in the Fluosol DA®-treated group.

No other products are in clinical trials for this indication.

5.2.5 Cancer Therapy

Oxygenation of hypoxic tumors during radiation or chemotherapy has been a long-term goal of radiologists and oncologists. PFC emulsions have been investigated in animal models for their ability to sensitize tumors to radiation and chemotherapeutics and consequently reduce the rate of tumor growth (Teicher and Rockwell 1983). Studies with oxygen electrodes have proven that animal tumors show an increased oxygen tissue tension when infused with PFCs while breathing carbogen (Song *et al.* 1987). Tumors appear to preferentially accumulate PFCs perhaps increasing their effectiveness as sensitizers by attracting macrophages to the tumor site (Long *et al.* 1978). Animal studies have shown that PFCs do not sensitize healthy tissues to radiation therapy (Mate and Rockwell 1985).

There have been at least six clinical trials of Fluosol DA® reported: three as an adjunct to radiation therapy and three as an adjunct to chemotherapy. The radiation studies involved head and neck, lung and glioma, while the chemotherapy studies treated non-small cell lung, colo/rectal and glioma.

The first trial (Rose *et al.* 1986) used Stage III/IV patients with head and neck cancer. These patients were given 7-9 mL/kg of Fluosol DA® (0.7 to 0.9 mL PFC/kg) on the first day of every week of therapy. They were given five weeks of therapy or 25 fractions of radiation and a total dose of 35 to 45 mL/kg of Fluosol DA® (3.5 to 4.5 ccPFC/kg). Patients breathed carbogen before and during each daily radiation treatment. There was some response to this therapy but the total number of patients was only 15. There were four cases of acute complement reactions, all of which were controlled with diphenhydramine. Eight of 15 patients exhibited serum enzyme elevations two to three times normal. These returned to normal three months post-therapy. Coagulation times, BUN, creatinine, serum albumin and bilirubin were all normal. White cell counts and hematocrit were slightly depressed, but these changes were attributed to the radiation. There was some evidence of increased incidence of mucositis in the Fluosol DA®-treated patients.

A larger trial of head and neck patients was completed by Lustig in 1989 (Lustig *et al.* 1989a). Forty-six patients were enrolled, and 37 completed the protocol. Eleven experienced complement side effects, and 17 of 46 exhibited serum enzyme elevations (ALT, AST and alkaline phosphatase) of two to five times normal. One-year survival is 67% compared to the RTOG average of 62%.

A Phase I/II study of non-small cell lung cancer patients treated with Fluosol DA® and oxygen was conducted in 1985 (Lustig *et al.* 1989b). Patients were enrolled in Stages II-IV without distant metastases. Fluosol dosing and radiation fractions were similar to the head and neck protocol discussed above with total doses of Fluosol DA® ranging from 42 to 49 mL/kg (4.2 - 4.9 mL/kg). Forty-five patients were enrolled, and 34 completed the program. There was a high frequency of complement responses in these patients, with 44% reacting to the test dose or showing post-infusion reactions. Typical abnormalities in serum enzymes were noted, with all returning to normal three to six months post therapy. Seventeen of 34 patients had a complete response, and 11 had a partial response. Thirteen patients remained alive 12 to 20 months post-treatment.

A Phase I/II trial of Fluosol DA® as an adjuvant to high-grade brain tumors has been reported (Evans *et al.* 1989). In this study, 18 patients were enrolled with a tumor grade of III to IV. Patients were given five to seven doses of 8 mL/kg of Fluosol DA®. Once again, a high percentage of the patients, 66%, experienced side effects upon infusion, and 11 of 18 patients had elevated serum enzyme levels had normalized three months post-treatment. Mean survival time in this small patient population was >64 weeks compared to historical survival on radiation therapy of 54 weeks.

Of the studies evaluating Fluosol DA® and oxygen breathing as an adjunct to chemotherapy, the most advanced application involves BCNU in high-grade glioma patients. The Phase I/II trial enrolled 51 patients who received between 150 and 600 mL/m^2 of Fluosol DA® and 200 mg/m^2 BCNU every six weeks for a total of three treatment regimes (Gruber *et al.* 1990). Of the 34 evaluable patients, 12 had a partial response (35%), 13 patients had stable disease and nine had disease progression. A Phase III trial is underway at the University of Kansas Medical Center now.

Oxygent CA, a 90 w/v% emulsion of PFOB, has been tested in animal models for adjuvant effect on cancer radiation therapy and chemotherapy with positive results (Teicher *et al.* 1992). No clinical data have been published.

Interstitial injection of an oxygenated, 35 v/v% emulsion of perfluorophenanthrene has been reported to cause a significant tumor-growth delay and increased survival in an animal model (Schweighardt and Woo

1988). The perceived advantage of this approach is that PFC is localized in the tumor, reducing the liver toxicity seen in all intravenously administered trials of Fluosol DA®. The limitation of this approach is that tumors must be palpable or otherwise visualizable to be injected. A Phase I study in head and neck cancer patients using this emulsion, trade-named Oncosol™, at doses up to 3 mL per tumor without observable toxicity. Additional clinical studies are being planned.

5.2.6 Imaging

PFC emulsions have been studied as contrast agents for x-ray, ultrasound and MRI, and as agents for direct ^{19}F MRI.

Long and coworkers investigated brominated PFCs, both neat and emulsified in the 1970's because of the known x-ray opacity of the bromine atom (Long *et al.* 1972a, 1972b, Long *et al.* 1978). They found that neat brominated PFCs could be administered intratrachially or orally and afforded excellent images without the side reactions. PFOB was subsequently developed by Alliance as a GI MRI contrast agent trade-named Imagent MR. PFOB contains no protons and as such appears as a dark void in MR images. Clinical trials demonstrated the efficacy of Imagent MR (Brown *et al.* 1991, Mattrey *et al.* 1991) in darkening the bowel and allowing the recognition of bowel from adjacent structures. In the Phase III clinical trial, 127 subjects were studied. They were imaged before and two to 60 minutes after they had ingested 500 to 1000 mL of PFOB. There were no acute or subacute GI symptoms. The product was rectally eliminated rapidly without uptake even by diseased mucosa. The product was approved by the FDA in 1993. Clinical usage has been limited, not for lack of efficacy, but due to expense. Alliance announced in September, 1994 that they would no longer support marketing of Imagent MR.

X-ray contrast with PFOB emulsions has been investigated for blood pool, liver and lymph node imaging. PFOB emulsions (100 w/v%) produced prolonged blood enhancement in animal models and because the particles were taken up by the macrophages of the liver and spleen, these organs also showed x-ray enhancement (Mattrey *et al.* 1984). Clinical trials have been conducted with this emulsion, trade-named Imagent BP (Behan *et al.* 1993, Bruneton *et al.* 1989). In the Behan study, the emulsion was administered to 18 cancer patients, 14 of whom had liver metastases. The emulsion doses ranged from 0.5 to 3.0 mL/kg. CT of the liver and spleen was performed before and immediately after infusion and again 24 hours later. PFOB increased the density of the blood, liver and spleen by 55, 39 and 317 Hounsfield units (HU), respectively. Tumor visualization was increased because the metastases enhanced minimally, 7 HU or lower, compared to the surrounding liver tissue. Peak enhancement of the liver and spleen occurred 24 hours post-infusion. Previously undetected metastases were found in two of the patients. Adverse events occurred in 14 of the 18 subjects. These included lower back pain, fever and malaise. The author concluded that the side effects could restrict the use of the product to a selected clinical population. In September of 1994, Alliance announced it would stop development of this product.

Ultrasound enhancement by perfluorocarbon emulsions has been extensively studied by Mattrey (Mattrey *et al.* 1982, Mattrey *et al.* 1983, Mattrey *et al.* 1987) using Fluosol DA® and perfluorooctyl bromide. These studies have shown enhancement of the liver, spleen, tumor, kidney and some of the vasculature. There has been one report of a clinical trial using Fluosol DA® as an ultrasound contrast agent (Mattrey *et al.* 1987). Administration of Fluosol DA® at 8 to 16 mL/kg caused rim or diffuse enhancement developed in liver metastases. Echogenic enhancement by Fluosol DA® allowed visualization of non-enhancing lesions which could not be seen before infusion. Side effects were minimal. Despite these seemingly positive results, there was no further clinical development of Fluosol DA® for ultrasound contrast enhancement. Alliance is currently developing an ultrasound contrast agent based on PFOB called Imagent US. This product is still in pre-clinical development.

Recently, Beppu and DeMaria reported the use of emulsions of low boiling point perfluorocarbons such as perfluoropentane which phase shifts from liquid at room temperature to gas particles when injected intravenously (Beppu *et al.* 1993, DeMaria *et al.* 1993). Because ultrasound contrast is dependent upon the density and particle size of the contrast media, these larger gas particles are inherently superior contrast agents compared to liquid emulsions. Pre-clinical studies in dogs with a PFP emulsion called QW3600 have generated excellent images. Unlike traditional microbubble contrast agents, QW3600 produces an intense and long lasting enhancement of the myocardium after intravenous injection. The myocardial contrast effect continued long after the effect had washed out of the ventricular chambers. Ligation of coronary arteries clearly defined the area at risk. Although, some loss of cardiac function was observed in this study. There was a dose-related decrease in arterial PO_2, increase in pulmonary artery pressure and decrease in cardiac output, blood pressure and dP/dt. The authors claimed these effects were not clinically significant. A Phase I clinical trial has been completed and at doses up to 0.1 ml/kg of emulsion there were no significant side effects (Cotter *et al.* 1994). Mild flushing and lightheadedness was observed in some of the subjects. These effects all resolved within two minutes. Dense and complete left ventricle opacification was seen at 0.1 ml/kg for an average of 2.8 minutes, but there was no myocardial opacification at this dose. The blood half-life was 2 minutes. PFC was found in the expired air for 10 to 16 minutes post-infusion.

The use of PFCs for MRI has a great deal of potential because of ^{19}F nucleus is magnetically active and highly sensitive. In addition, there is only trace naturally occurring ^{19}F, so there is no natural background signal to interfere with the diagnostic signal. Early attempts to use ^{19}F imaging were successful but not clinically useful (Joseph *et al.* 1985, McFarland *et al.* 1985, Longmaid *et al.* 1985) because most PFCs have multiple signals causing misregistration of signals and diluting the amount of ^{19}F signal per molecule infused. Confounding this is the short T_2 time of most PFCs (< 6 milliseconds). Thus, the signal is not only dilute but short lived.

Schweighardt cleverly solved the problem of multiple signals by using perfluoro-15-crown-5 ether (PFCE) in which all 22 of the fluorine atoms are identical and form a single peak (Schweighardt 1989). In addition, the T_2 relaxation time of PFCE is 200 milliseconds compared to 6 milliseconds for PFOB and PFDCO. Using a 40 v/v% emulsion of PFCE, excellent liver, spleen, tumor and vascular images have been obtained at doses of 3 mL PFC/kg in animals (Dardzinski and Sotak 1994).

Clark and Ackerman were the first to discover that the T_1 relaxation times of PFCs change with oxygen tension (Clark *et al.* 1985). They postulated that the dependence of T_1 on oxygen tension could be used to construct oxygen maps of the tissues in which the PFC was in contact. McFarland reduced this concept to practice in cats where 30% of the blood was replaced with an emulsion of perfluorotripropyl amine (PFTPA), and oxygen maps of the brain were obtained (Eidelberg *et al.* 1988). However, the large dose required coupled with the software modifications required to suppress the misregistered signals precluded the clinical utility of this emulsion. Sotak has recently used emulsions of PFCE to construct oxygen maps of the liver, spleen and tumors in mice. When animals were treated with 3 mL PFC/kg of PFCE, significant differences in oxygen tension could be measured in the liver, spleen and tumors of mice when they were breathing air compared to carbogen (Dardzinski and Sotak 1994).

5.2.7 Liquid Breathing

Since Clark's initial liquid breathing experiment, most of the research on PFCs has focused on the development of emulsified forms suitable for intravenous administration. However, a small but steady effort to establish liquid breathing as a therapeutic for lung lavage and respiratory distress has been continuing. Modell and coworkers worked out the mechanics of long term liquid breathing of animals for evaluation of PFC toxicity, adsorption and pulmonary physiology (Modell, Newby and Ruiz 1970, Modell *et al.* 1976). Shaffer has made significant progress in both the physiology of liquid breathing and in developing respirators adapted for liquid breathing (Shaffer and Moskowitz 1974). This work culminated in the first human clinical trial in pre-term neonates. A 19-week old neonate with respiratory distress syndrome was instilled with 30 mL/kg of an oxygenated PFC. Liquid ventilation was initiated as two 3 -

5 minute cycles followed by conventional ventilation. During liquid ventilation cycles tidal volumes of 15 mL/kg were delivered (Greenspan *et al.* 1990). The infant survived for 19 hours with markedly improved pulmonary parameters and expired from other causes.

Alliance has conducted a Phase I study in similar neonates using PFOB (LiquiVent™). In this protocol, the infants had exhausted conventional therapy including two regimes of surfactant. The infants were liquid ventilated with PFOB for 15 minutes. Of seven neonates treated, two survived (Hopkins 1994). Alliance announced in October, 1994 that they would begin enrolling patients for a Phase I study of liquid ventilation in adult respiratory distress syndrome (ARDS).

5.3 Conclusions

Significant progress has been made in many areas in the field of medical perfluorocarbon research. Many of the problems which prevented the use of PFC emulsions for oxygen transport have been overcome, but limitations will still be encountered at some dose because of thrombocytopenia and bioaccumulation. Research to solve these problems is underway. The use of PFCs in MI failed in a large clinical trial, and there seems to be no compelling reason to pursue additional formulations for this indication. Cancer therapy still holds great promise but has been set back by the withdrawal of the leading company in the field. Advances in the use of PFCs for ultrasound contrast and direct ^{19}F MRI have been made, but other imaging uses such as CT and MRI contrast seem to have toxicity, efficacy or cost limitations. Liquid breathing has seen an exciting rejuvenation due to the results in neonates, however, it remains to be seen if the far more complex ARDS can be successfully treated with liquid ventilation.

5.4 References

Bajaj, A.K., M.A. Cobb, R. Virmani, *et al.* Limitation of myocardial reperfusion injury by intravenous perfluorochemicals. *Circulation* 79: 645-656, 1989.

Behan, M., D.O. O'Connell, R.F. Mattrey, and D.N. Carney. Perfluoroctyl bromide as a contrast agent for CT and sonography: preliminary clinical results. *Am. J. Radiol.* 160: 399-405, 1993.

Beppu, S., H. Matsuda, T. Shishido, *et al.* Success of myocardial contrast echocardiography by peripheral venous injection method: visualization of area at risk. *Circulation* 88: I-401 [abstract], 1993.

Blauth, C.I., P.L. Smith, J.V. Arnold, *et al.* Influence of oxygenator type on the prevalence and extent of microembolic retinal ischemia during cardiopulmonary bypass. *J. Thorac. Cardiovasc. Surg.* 99: 61-69, 1990.

Brown, J.J., J.R. Duncan, J.P. Heiken, *et al.* Perfluorooctyl bromide as a gastrointestinal contrast agent for MR imaging: use with and without glucagon. *Radiobiology* 181: 455-460, 1991.

Bruneton, J.N., M.N. Falawese, *et al.* Liver, spleen and vessels: preliminary clinical results of CT with perfluorooctyl bromide. *Radiology* 170: 179-183, 1989.

Cernaianu, A.C., R.K. Spence, T. Vasilidze, *et al.* Transfusion triggers with perflubron (Oxygent™) in a canine model of surgical hemodilution. In: *Fifth International Symposium on Blood Substitutes: New Frontiers* (T. Chang, J.G. Riess and R.M. Winslow, eds.) F3 [abstract], 1993.

Clark, L.C., J. Ackerman, S. Thomas, *et al.* Perfluorinated organic liquids and emulsions as biocompatible NMR imaging agents for ^{19}F and dissolved oxygen. *Adv. Exp. Med. Biol.* 180: 835-846, 1985.

Clark, L.C., and F. Gollan. Survival of mammals breathing organic liquids equilibrated with oxygen at atmospheric pressure. *Science* 152: 1755-1756, 1966.

Cleman, M., C.C. Jaffe, and D. Wholgelernter. Prevention of ischemia during percutaneous transluminal coronary angioplasty by transcatheter infusion of oxygenated Fluosol DA-20%. *Circulation* 74: 555-562, 1986.

Cotter, B.O. L. Kwan, B. Kimura, *et al.* Evaluation of the efficacy, safety and pharmacokinetics of QW3600 (Echogen) in man. *Circulation* 90: I-555, 1994.

Dardzinski, B.J., and C.H. Sotak. Rapid tissue oxygen tension mapping using ^{19}F inversion recovery echo planar imaging of perfluoro-15-crown-5-ether. *Mag. Reson. Med.* 32: 88-97, 1994.

DeMaria, A.N., H. Dittrich, O.L. Kwan, *et al.* Myocardial opacification produced by peripheral venous injection of a new ultrasonic contrast agent. *Circulation* 88: I-401 [abstract], 1993.

Eidelberg, D., G. Johnson, D. Barnes, *et al.* ^{19}F NMR imaging of blood oxygenation in the brain. *Mag. Reson. Med.* 6: 344-352, 1988.

Evans, R.G., B.F. Kimler, R.A. Morantz, and R.A. Batnitzky. Lack of complications in long-term survivors after treatment with Fluosol® and oxygen as an adjuvant to radiation therapy for high-grade glioma. *Int. J. Rad. Oncol. Bio. Phys.* 26: 649-652, 1989.

Flaim, S., D.R. Hazard, J. Hogan, and R.M. Peters. Characterization and mechanism of side-effects of Imagent BP (highly concentrated fluorocarbon emulsion) in swine. *J. Invest. Radiol.* 26: S122-S124, 1991.

Geyer, R.P., R.C. Monroe, and K. Taylor. Survival of rats totally perfused with a fluorocarbon-detergent preparation. In *Organ Perfusion and Preservation* (J. Norman, ed.). New York: Appleton Century and Crofts, 1968, pp. 85-97.

Glogar, D.H., R.A. Kloner, J. Muller, *et al.* Fluorocarbons reduce myocardial ischemic damage after coronary occlusion. *Science* 211: 1439-1441, 1981.

Goodin, T.H., E.G. Grossbard, R.J. Kaufman, *et al.* A perfluorochemical emulsion for prehospital resuscitatien of experimental hemorrhagic shock: a prospective, randomized controlled study. *Crit. Care Med.* 22: 680-689, 1994.

Gould, S.A., A.L. Rosen, L.R. Sehgal, *et al.* Fluosol DA as a red-cell substitute in acute anemia. *N. Eng. J. Med.* 314: 1653-1656, 1986.

Greenspan, J.S., M.R. Wolfson, D. Rubenstein, and T.H. Shaffer. Liquid ventilation of human preterm neonates *J. Pediatrics* 117: 106-111, 1990.

Gruber, M., M. Prados, C. Russell, *et al.* Phase I/II study of Fluosol® and oxygen in combination with BCNU in malignant glioma. *Proc. Am. Assoc. Cancer Res.* 31: 190, 1990.

Hopkins, R. Personal communication, 1994.

Jaffe, C.C., D. Wohlgelernter, H. Cabin, *et al.* Preservation of left ventricular ejection fraction during percutaneous transluminal coronary angioplasty by distal transcatheter coronary perfusion of oxygenated Fluosol DA 20%. *Am. Heart J.* 6: 1156-1164, 1988.

Joseph, P.M., Y. Yuasa, H.L. Kundel, *et al.* Magnetic resonance imaging of fluorine in rats infused with artificial blood. *Invest. Radiol.* 20: 504-509, 1985

Kaufman, R.J. Medical oxygen transport using perfluorochemicals. In: *Biotechnology of Blood* (J. Goldstein, ed.). Boston: Butterworth-Heinemann, 1991, pp. 127-162.

Kaufman, R.J. Perfluorochemical emulsions as blood substitutes. In *Emulsions, a Fundamental and Practical Approach* (J. Sjoblom, ed.). Boston: Kluwer Academic Publishers, 1992, pp. 207-226.

Kaufman, R.J. The results of a Phase I clinical trial of a 40 v/v% emulsion of HM351 (Oxyfluor™) in healthy human volunteers. *Artif. Cells, Blood Substitutes, Immobil. Biotech.* 22: A112, 1994.

Keipert, P. Use of Oxygent™, a perfluorochemical (PFC) emulsion, as an antihypoxic agent to improve tissue oxygenation during acute blood loss. In: *Blood Substitutes and Related Products: Advances in Development, Trial Design and Clinical Development* (E. Scatchard and M. Mc-

Bride, eds.) Southborough, MA: International Business Communications, 1994.

Long, D.M., M.S. Liu, P.S. Szanto, *et al*. Efficacy and toxicity studies with radioopaque perfluorocarbon. *Radiology* 133: 71, 1972a.

Long, D.M., M.S. Liu, P.S. Szanto, *et al*. Forefront: preliminary communication - initial observations with a new x-ray contrast agent radioopaque perfluorocarbon. *Rev. Surg.* 29: 71, 1972b.

Long, D.M., F.K. Multer, A.G. Greenburg, *et al*. Tumor imaging with x-rays using macrophage uptake of radioopaque fluorocarbon emulsions. *Surgery* 84: 104-112, 1978.

Longmaid, H., D. Adams, *et al*. In vivo ^{19}F NMR imaging of liver, tumor and abcesses in rats. *Invest. Radiol.* 20: 141-145, 1985.

Lustig, R., N. McIntosh-Lowe, C. Rose, *et al*. Phase I/II study of Fluosol-DA and 100% oxygen as an adjuvant to radiation in the treatment of advanced squamous cell tumors of the head and neck. *Int. J. Rad. Oncol. Biol. Phys.* 16: 1587-1593, 1989a.

Lustig, R., N. Lowe, L. Prosnitz, *et al*. Phase I/II study of Fluosol and 100% oxygen breathing as an adjuvant to radiation in the treatment of unresectable non-small cell carcinoma of the lung. *Int. J. Rad. Oncol. Biol. Phys.* 17: 202, 1989b.

Mate, T.P., and S. Rockwell. Perfluorochemical emulsions do not affect bone marrow radiosensitivity. In: *Abstracts, American Society Therapy Radiation Oncologists Meeting*, Washington, D.C., Oct. 1984.

Mattrey, R.F., F.W. Scheible, B.B. Gosink, *et al*. Perfluorooctyl bromide: a liver and spleen specific and a tumor imaging ultrasound contrast material. *Radiology* 145: 759-762, 1982.

Mattrey, R.F., G.R. Leopold, E. vanSonneberg, *et al*. Perfluorochemicals as a liver and spleen seeking ultrasound contrast agent. *J. Ultrasound Med.* 2: 173-176, 1983.

Mattrey, R.F., D.M. Long, R.A. Slutsky, and G.B. Higgins. Perfluorooctyl bromide as a blood pool contrast agent for liver, spleen and vascular imaging. *J. Comput. Assist. Tomogr.* 8: 739-744, 1984.

Mattrey, R.F., G. Strich, R.E. Shelton, *et al*. Perfluorochemicals as ultrasound contrast agents for tumor imaging and hepatosplenography: preliminary clinical results. *Radiology* 163: 339-343, 1987.

Mattrey R.F., M.A. Trammert, J.J. Brown, *et al*. Oral contrast agents for magnetic resonance imaging. *Invest. Radiol.* 26: S65-S66, 1991.

McFarland, E.R., J.A. Koucher, B.R. Rosen, *et al*. In vivo ^{19}F NMR imaging. *J. Comp. Assist. Tomagr.* 9: 8-15, 1985.

Modell, J.H., E.J. Newby, and B.C. Ruiz. Long-term survival of dogs after breathing oxygenated fluorocarbon liquid. *Fed. Proc.* 34: 312-320, 1970.

Modell, J.H., H.W. Calderwood, B.C. Ruiz, *et al.* Liquid ventilation of primates. *Chest* 69: 79-81, 1976.

Moody, D.M., M.A. Bell, V.R. Challa, *et al.* Brain microemboli during cardiac surgery or aortography. *Annu. Neurol.* 28: 477-486, 1990.

Nunn, G.R., G. Dance, J. Peters, and L.H. Cohn. Effect of fluorocarbon exchange transfusion on myocardial infarction in dogs. *Am. J. Cardiol.* 52: 203-205, 1983.

Riess, J.G., and M. Leblanc. Perfluoro compounds as blood substitutes. *Angewande Chemie International Edition in English* 17: 621-634, 1978.

Riess, J.G. Reassessment of the criteria for the selection of perfluorochemicals for second generation blood substitutes: analysis of structure-property relationships. *Artif. Organs* 8: 44-56, 1984.

Rose, C.M., R. Lustig, N. McIntosh, and B.A. Teicher. A clinical trial of Fluosol DA 20% in advanced squamous cell carcinoma of the head and neck. *Int. J. Rad. Oncol. Biol. Phys.* 12: 1325-1327, 1986.

Schweighardt, F.K., and D. Woo. Interstitial administration of perfluorochemical emulsions for reoxygenation of hypoxic tumor cells. *US Patent* No. 4,781,676, 1988.

Schweighardt, F.K. Perfluoro-crown ethers in fluorine magnetic resonance imaging. *US Patent* No. 4,838,274, 1989.

Shaffer, T.H., and G.D. Moskowitz. Demand-controlled liquid ventilation of the lungs, *J. Appl. Physiol.* 36: 208-213, 1974.

Shaw P.J., D. Bates, N.E. Cartlidge, *et al.* Neurologic and neuropsychological morbidity following major surgery: comparison of coronary bypass and peripheral vascular surgery. *Stroke* 18: 700-707, 1987.

Sloviter, H., and T. Kamimoto. Erythrocyte substitute for perfusion of brain. *Nature* 216: 458, 1967.

Song, C.W., I. Lee, T. Kasegawa, *et al.* Increase in PO_2 and radiosensitivity of tumors by Fluosol-DA(20%). *Cancer Res.* 47: 442-446, 1987.

Spears, R., J. Serur, D. Baim, *et al.* Myocardial protection with Fluosol-DA during prolonged coronary balloon occlusion in the dog. *Circulation* 73: II-245 [abstract], 1983.

Spence, R.K. Use of Oxygent^{TM}, perfluorocarbons in the twenty-first century: clinical applications as transfusion alternatives. In *Blood Sub-*

stitutes and Related Products (E. Scatchard and M. McBride, eds.) Southborough, MA: International Business Communications, 1994.

Spiess, B.D., R.J. McCarthy, K.J. Tuman, A.W. Woronowicz, K.A. Tool, and A.D. Ivanovich. Treatment of decompression sickness with a perfluorocarbon emulsion (FC-43). *Undersea Biomed. Res.* 15: 31-37, 1988.

Taylor, K.M., J.V. Arnold, J. Fleming, A.C. Bird, E.G. Grossbard, and R.J. Kaufman. Cerebral protection during CPB using perfluorocarbon: a preliminary report. In *Second International Conference on the Brain and Cardiac Surgery.* Key West FL [abstract], 1992.

Teicher, B.A., and S. Rockwell. Increased efficacy of radiotherapy in mice treated with perfluorochemical emulsions plus oxygen. *Am. Assoc. of Cancer Res.,* Abstract 25-28, 1983.

Teicher, B.A., S.A. Holden, G. Ara, et al. A new concentrated perfluorochemical emulsion and carbogen breathing as a adjuvant to treatment with antitumor alkylating agents. *J. Cancer Res. Clin. Oncol.* 118: 509-514, 1992.

Tremper, K.K., G.M. Vercellotti, and D.E. Hammerschmidt. Hemodynamic profile of adverse clinical reactions to Fluosol DA 20%. *Crit. Care Med.* 12: 428-431, 1984.

Wall, T.C., R.M. Califf, J. Blamkenship, et al. Intravenous Fluosol in the treatment of acute myocardial infarction. *Circulation* 90: 114-126, 1994.

Chapter 6

Design of Chemically Modified and Recombinant Hemoglobins as Potential Red Cell Substitutes

James M. Manning, Ph.D.

Rockefeller University, 1230 York Avenue, New York, New York   10021

6.1 Introduction

Modified cell-free hemoglobin derivatives have been under study as potential red cell substitutes for nearly three decades (Winslow 1992). These derivatives are prepared by treatment of hemoglobin with some chemical reagent to yield a product that possesses a property considered desirable for a blood substitute. The degree of sophistication of these derivatives has, in general, increased in proportion to our knowledge of hemoglobin itself.

More recently, recombinant DNA technology has been developed for hemoglobin in general and for blood substitute research in particular. At present, both the chemical and the recombinant DNA approaches are being evaluated in order to determine which is the more preferable route to a practical, clinically useful and non-toxic blood substitute. It is quite possible that some combination of these will eventually be used. Each has advantages over the other since in some instances their capabilities are different. In this chapter, these approaches will be treated separately with some emphasis on studies from the author's laboratory.

6.2 Allosteric Modifiers of Hemoglobin Function

6.2.1 Covalent Chemical Modifiers

The strategy that some investigators have used for developing a useful blood substitute has been to imitate the effects of the natural allosteric

Blood Substitutes: Physiological Basis of Efficacy
Winslow et al., Editors
© Birkhäuser Boston 1995

regulators of oxygen binding to hemoglobin in the red cell - chloride, carbon dioxide and 2,3-diphosphoglycerate (DPG) in shifting the oxygen equilibrium curve of the cell-free purified hemoglobin to the right (Figure 6.1).

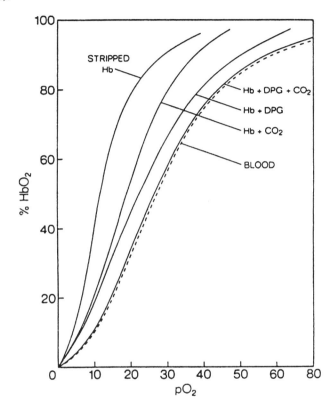

Figure 6.1 From Kilmartin and Rossi-Bernardi, *Physiol. Rev.* 53: 836, 1973.

However, since these natural regulators bind reversibly to hemoglobin, they are of little value in themselves in attaining the goals of a hemoglobin-based red cell substitute because in the plasma (outside the red cell) they would diffuse from the hemoglobin. Some investigators use the strategy of first identifying the binding sites of these natural regulatory molecules and then performing covalent chemistry at these same sites in order to mimic the desired physiological effect. The type of chemical modification described in this section is the traditional selective (or specific) chemical modification rather than the random type of chemical modification described later in this article. An advantage in studying covalent agents is that the exact site of modification can be easily determined. Such information can sometimes permit identification of a particular site on the protein that controls a normal physiological function of hemoglobin. The mode of action of a covalent agent is dictated by the site(s) to which the modifier is attached, the chemical nature of the adduct, and the parts of the hemoglobin molecule with which the covalent moiety interacts. The extensive body of knowledge on hemoglobin usually permits an explanation of its mechanism of action (Manning 1991).

In order for the agent to be effective, it should have a high degree of specificity, which means that the chemical modification reaction is more efficient with one particular site on hemoglobin than it is with the same type of amino acid side chain on other parts of hemoglobin or on other proteins. This is especially important considering the bulk of hemoglobin that would likely be administered for infusion of a hemoglobin-based red cell substitute. Such specificity could arise either from the intrinsic reactivity of the functional group itself or from its environment in the protein which may lead to an enhanced binding of the reagent. In general, success in interpreting the effects of a particular modification on the functional properties of the hemoglobin molecule is more likely if the sites of modification are limited or, even better, restricted to one particular amino acid residue. Covalent modifiers have advantages over the noncovalent type since, in general, they are effective at low concentrations and the adducts do not diffuse from the protein. Many of the covalent agents used in blood-substitute research have amino groups as their targets since this class of functional groups has proved to be among the most amenable to chemical modifications.

6.2.2 Binding of CO_2 to Hemoglobin

The strategy of identifying the site(s) of interaction of the natural, reversibly-bound allosteric regulator, carbon dioxide, and then modifying this same site covalently in order to attain a permanently lowered oxygen affinity is illustrated here. There were some early suggestions that the N-terminal amino groups of the α- and β-chains were major contributors to the binding and physiological effects of CO_2 in lowering the oxygen affinity (Perrella, Rossi-Bernardi and Roughton 1972). In attempts to identify these binding sites unambiguously, we used a stable analog of CO_2 binding, the carboxymethyl (Cm) group, which was introduced at the N-terminal amino groups under mild conditions at neutral pH by treatment of the hemoglobin with sodium glyoxylate in the presence of sodium cyanoborohydride (DiDonato *et al.* 1983, Fantl *et al.* 1987a, Fantl *et al.* 1987b).

N-Carboxymethylation

$$\text{Hb-NH}_2 + \underset{\underset{\text{COOH}}{|}}{\text{CHO}} \xrightarrow{\text{NaCNBH}_3} \text{Hb-NH-CH}_2\text{COOH}$$

Pilot studies were performed to establish the ratio of reactant to protein, the optimum pH of the reaction, and the time of incubation. Peptide mapping, amino acid analysis, and x-ray diffraction analysis of the purified carboxymethylated hemoglobin product indicated that the site of reaction at the N-terminal valine of both chains of hemoglobin was the same site that bound CO_2. Another proof that this derivative did indeed behave as a CO_2 analog was the finding that the hemoglobin tetramer carboxymethylated on all four N-terminal residues, $\alpha_2^{Cm}\beta_2^{Cm}$, had a very low oxygen affinity (P50 = 37 mm Hg), which is the same effect that CO_2 has on hemoglobin (Fantl *et al.* 1987a,b).

6.2.3 Binding of Chloride to Hemoglobin

The Rome hemoglobin research team of Chiancone, Antonini, Brunori and their colleagues demonstrated specific binding of chloride to particular sites on the hemoglobin molecule (Chiancone *et al.* 1972). Studies in our laboratory showed that the N-terminus of the α-chain was one of the major chloride binding sites (Nigen, Manning and Alben 1980). A second major site of chloride binding was established by several laboratories as the region that binds 2,3-DPG in a cleft between the two β-chains also in the deoxy conformation (Arnone 1972, Bonaventura *et al.* 1976).

By the early 1980's when these two sites were established as being major contributors for functional chloride binding, many investigators realized that there still remained a residual 20-25% chloride-induced lowering of the oxygen affinity that was not accounted for by these two sites. This residual effect is shown in Figure 6.2 for the CO_2 binding sites.

Using the carboxymethyl derivative for the CO_2 binding reaction (a covalent anion that had a charge like that of the chloride anion), the results showed that when the two sites for CO_2 binding were blocked $(\alpha_2^{Cm}\beta_2^{Cm})$, the residual chloride effect also amounted to 20-25%, *i.e.*, the slope of the line in the bottom panel compared to the slopes in the three panels above (DiDonato *et al.* 1983).

It was possible that a number of distinct sites, each making a small chloride contribution, accounted for this remaining 20-25%, and that specific chemical modifications by their very nature of purifying a product might miss this site(s). Hence, a new procedure, which is referred to as regiorandom chemical modification, was employed for their identification (Ueno and Manning 1992, Ueno, Popowicz and Manning 1994). Random chemical modification involved analysis of a mixture of peptides after performing the reaction under very mild conditions so there would not be extensive modification. It was important to have a reagent that would bind to the chloride binding sites, yet not cause disruption of the protein. A bifunctional acetylating agent, methyl acetyl phosphate (MAP), was employed. Independent studies show that this reagent had the same effects on the functional properties of bovine hemoglobin in terms of a lower P50, as did chloride. Furthermore, the effects were mutually competitive. Thus, there were several lines of evidence that the methyl acetyl phosphate reagent could be used with confidence as a probe with chloride binding sites.

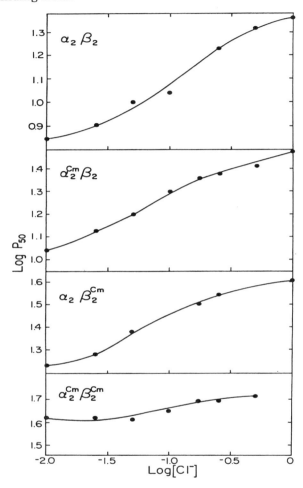

Figure 6.2 Effect of chloride on the oxygen affinity of specifically carboxymethylated hemoglobin.

By this random modification approach, two functional chloride binding sites per α-chain and three functional sites per β-chain were identified. Some of these comprised the two well-known regions described above and the Lys-99(α) previously identified as a functional chloride binding site (Vandegriff *et al.* 1989). Other sites were new. Molecular modeling showed the relationship of this other chloride binding region to the other two major sites. Unexpectedly, and as described below, the location of this third chloride binding region is very much related to current blood-substitute research.

This third functional chloride binding region was found in the central dyad axis of bovine hemoglobin, which is a focus of much current research on hemoglobin-based red cell substitutes. When viewed through the central dyad axis of the molecule, a relationship between all three sites becomes apparent, *i.e.,* the sites within the central axis, Lys-99(α) and Lys-103(β), are aligned along the sides connecting the major oxygen-linked sites which appears to form a channel connecting them (Figure 6.3). Furthermore, there appears to be a symmetrical relationship between these functional chloride binding sites in deoxyhemoglobin that disappears upon oxygenation. It is possible that the binding of chloride at these sites helps to hold the central dyad axis in its more open configuration by repulsion of the negatively-charged chloride ions. Rather than the previously held notion that there are specific chloride binding sites, it is more correct to view them as chloride binding regions. Indeed, the regio-random approach of chemical modification was designed to identify such sites. This suggestion has also been advanced by Perutz, Shih and Williamson (1994).

The working hypothesis is that any constraint that prevents the constriction of the dyad axis leads to a low oxygen affinity of hemoglobin. For example, the introduction of a covalent cross-link between the two lysine-99(α) side chains across the central dyad axis in the deoxy conformation (Chatterjee *et al.* 1986, Vandegriff *et al.* 1989), leads to a significantly lower oxygen affinity. This result is also in agreement with the findings of Lalezari *et al.* (1990) and Abraham *et al.* (1992) who found that large organic molecules that lower the oxygen affinity of hemoglobin also bound in this region.

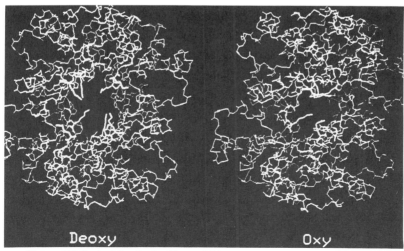

Figure 6.3 Central dyad axis of hemoglobin.

6.2.4 Overlapping Functions of Allosteric Regulators

There are indications in the literature that there may be some overlapping (or redundant) regions of the allosteric regulators chloride and CO_2. Indeed, the data in Figure 6.2 are consistent with this concept. The findings of Winslow and Vandegriff and their colleagues (1989), who reported that the $\alpha\alpha$-cross-linked hemoglobin derivative lost a large fraction of its chloride-induced decrease in oxygen affinity, represent convincing data for the idea of overlapping regions for allosteric regulators. Apparently, the cross-link across the dyad axis between the two Lys-99(α) side chains is of such strength due to its covalent nature that it achieves a very large effect in lowering the oxygen affinity in the absence of chloride. The suggestion of contiguous chloride binding regions (Figure 6.3) actually provides a framework for further understanding of this concept.

6.3 Cross-linking of Hemoglobin Subunits

Various derivatives of pyridoxal have been used in order to lower the oxygen affinity of hemoglobin. Their effects have been summarized by Winslow (1992). Among these were various derivatives of pyridoxal 5'-phosphate, which were used to mimic the effect of 2,3-DPG in lowering the oxygen affinity (Benesch and Benesch 1981).

6.3.1 Cross-links between α Chains

More emphasis has been placed on chemical reagents that cross-link hemoglobin rather than on the monofunctional type of compounds described in the paragraph above. The reason for this is that tetrameric hemoglobin is in a dynamic equilibrium with its constituent $\alpha\beta$ dimers.

$$\alpha_2\beta_2 \rightleftharpoons 2\alpha\beta$$

Practically all the hemoglobin in the red cell is in the tetrameric state. However, with circulating cell-free hemoglobin, such as a blood substitute, the presence of such dimers from the above equilibrium would lead to blockage of the glomeruli of the kidney and the continuous drainage of the infused tetrameric hemoglobin from the circulation. Therefore, cross-linking is essential to prevent this clearance as first pointed out by Bunn and Jandl (1968). An example of the importance of this cross-linking reaction are the studies on the plasma retention times, which were carried out in collaboration with Hess, Winslow and colleagues when they were at the Letterman Army Institute of Research (Manning 1991). The hemoglobin derivative with a cross-link between the N-terminal amino acids of its α-chains is referred to as DIBS-Hb (DIBS is diisothiocyanato benzene sulfonate). This preferred site of cross-linking comes about because the size of the cross-link is well accommodated at this site in comparison to other possible sites, although there is a small amount of reaction at other sites.

Studies on Cm hemoglobin, which had a lowered oxygen affinity (P50 = 37 mm Hg) as described above, indicated a plasma retention time of between 35 and 40 minutes in the circulation of rats, a value similar to the retention time of unmodified hemoglobin A (Hess *et al.* 1989). In contrast, the DIBS-cross-linked hemoglobin tetramer had a half-life of about 3 hours (Figure 6.4). However, this derivative did not have a low oxygen affinity. These results showed that the increased plasma retention time was a function of the lack of ability to dissociate into tetramers and not its oxygen affinity. Further studies in progress are aimed at determining the plasma retention time of highly purified and defined hemoglobin cross-linked species of 128,000 and 256,000 Dalton molecular weight.

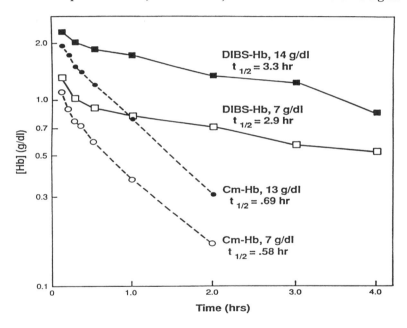

Figure 6.4 Plasma retention time of Cm-Hb and DIBS hemoglobin.

6.3.2 Other Cross-linking Agents

Various other derivatives of pyridoxal have been used to cross-link hemoglobin subunits within a tetramer (Benesch *et al.* 1972, 1975, 1981, 1984, Benesch, Benesch and Kwong 1982, Benesch and Kwong 1988). Glutaraldehyde was used in the early studies of hemoglobin cross-linking but it is a rather non-specific agent and yields mainly high molecular weight products, which in some instances may be desirable. Another useful cross-linking agent, because it reacts slowly and, therefore, the progress of the reaction can be controlled, is glycolaldehyde (Manning and Manning 1988). Others include various cross-links between the β-chains (Bucci *et al.* 1989) and detran-hemoglobin cross-links (Chang and Geyer 1988; Tam, Blumenstein and Wong 1976). The acyl phosphates are also good cross-linking agents (Kluger *et al.* 1992). The combination of pyridoxal modification and polymerization has also been used (DeVenuto and Zegna 1982; Moss *et al.* 1989).

6.4 Recombinant Hemoglobins

6.4.1 Recombinant Hemoglobin in Bacteria

The first useful procedure for expressing recombinant hemoglobin was reported in *E. coli* by expressing a globin cDNA fused to the coding region of a bacteria phage repressor protein (Nagai and Thøgersen 1984). The strategy was that the fusion protein contained a sequence that permitted enzymatic digestion to give the correct N-terminus. The purified globin was subsequently reconstituted with heme to produce a functional hemoglobin. Several improvements of this original procedure have been published subsequently. Since *E. coli* is a prokaryotic system, the processing of the nascent protein is different from that of human hemoglobin synthesized in reticulocytes. Thus, in the bacterial systems, there is usually a methionine at the N-terminus. This problem was recently overcome by the introduction of the gene encoding the enzyme methionine aminopeptidase on the same plasmid (Shen *et al.* 1993). In general, the bacterial expression system gives a reasonably high yield of hemoglobin.

6.4.2 Expression in Yeast

Expression in the eukaryote *S. cerevisiae* was designed with the objective of avoiding fusion proteins (Wagenbach *et al.* 1991). A synthetic α and β-globin cDNA sequence was incorporated on a single plasmid with the yeast promoters. The system produced equal amounts of α and β-globin chains and utilized the endogenous yeast heme to produce a soluble hemoglobin tetramer. The expression of the hemoglobin in yeast was somewhat lower than that in *E. coli* but had the advantage that the protein was processed and folded correctly (Martin de Llano *et al.* 1993a,b). This was shown by a number of biochemical and physiological techniques for standard hemoglobins, such as HbA and HbS.

This system was used to express the recombinant Hb where Asn-108(β) was replaced by a Lys (Wagenbach *et al.* 1991). This recombinant mutant Hb, which when found in nature is called Hb Presbyterian, has a low oxygen affinity possibly because a new chloride binding site has been introduced. It turns out that the site of this substitution is within the central dyad axis. It is possible that the introduction of a new chloride binding site in this region holds the hemoglobin tetramer in the more open deoxy conformation, similar to the $\alpha\alpha$-cross-linked Hb described above.

A recombinant mutant hemoglobin with Asn-102(β) replaced by an Ala [called N102A(β)] was recently prepared using this yeast expression system (Yanase *et al.*, in press). The side chain of Asn-102(β) is part of an important region of the $\alpha_1\beta_2$ interface that undergoes large structural changes in the transition between the deoxy and oxy conformations. Alanine was chosen as the replacement because its methyl side chain cannot participate in hydrogen bond formation thought to be important at this site, yet it is small enough not to disrupt the subunit interface. Its oxygen binding curve indicated a very high P50 of 42 mm Hg in the absence of chloride. In the presence of added chloride, its oxygen affinity was further reduced only slightly to a P50 of 49 mm Hg. Thus, the results with N102A(β) indicate that the maximal decrease in oxygen affinity may already have been achieved and that chloride is not necessary.

6.4.3 Expression in Insect Cells

There has been some success in expressing hemoglobin in insect cells infected with the baculovirus (Groebe *et al.* 1992). In some cases, denatured α and β-globin chains accumulated in these cells since heme, although present, was not incorporated into the chains. Also, there was poor expression of the α-globin gene in this system. The future of this approach is uncertain at present because it requires some modification.

6.4.4 Expression in Transgenic Animals

The initial reports on the expression of hemoglobin in the transgenic mouse led to enthusiasm about incorporating other genes in this system to study various disease processes. Typical was the production of the transgenic mouse for sickle cell anemia. Subsequently, a transgenic pig system, where there was the possibility of large amounts of hemoglobin, was achieved (Kumar, in press). This system was successful in the transfer of genetic material in offspring to pigs that were producing both human hemoglobin A and pig hemoglobin. Such a system obviously has the advantage that the large amounts of material required for blood-substitute research would be available. In addition, the pig reticulocyte processes hemoglobin just like the human red cells do.

6.5 Acknowledgments

Supported in part by NIH Grant HL-48018 and U.S. Army Contract DAMD17-94-V-4010.

6.6 References

Abraham, D.J., F.C. Wireko, R.S. Randad, C. Poyart, J. Kister, B. Bohn, J.-F. Liard, and M.P. Kunert. Allosteric modifiers of hemoglobin: 2-[4-[[(3,5-disubstituted anilino) carbonyl]methyl]phenoxy]-2-methylpropionic acid derivatives that lower the oxygen affinity of hemoglobin in red cell suspension, in while blood, and *in vivo* in rats. *Biochemistry* 31: 9141-9149, 1992.

Arnone, A. X-ray diffraction study of binding of 2,3-diphosphoglycerate to human deoxyhaemoglobin. *Nature* 237: 146-149, 1972.

Benesch, R., and R.E. Benesch. Preparation and properties of hemoglobin modified with derivatives of pyridoxal. *Meth. Enzymol.* 76: 147-158, 1981.

Benesch, R., R.E. Benesch, and S. Kwong. Labeling of hemoglobin with pyridoxal phosphate. *J. Biol. Chem.* 257: 1320-1324, 1982.

Benesch, R.E., R. Benesch, R.D. Renthal, and N. Maeda. Affinity labeling of the polyphosphate binding site of hemoglobin. *Biochemistry* 11: 3576-3582, 1972.

Benesch, R., R.E. Benesch, S. Yung, and R. Edalji. Hemoglobin covalently bridges across the polyphosphate binding site. *Biochem. Biophys. Res. Comm.* 63: 1123-1129, 1975.

Benesch, R.E., and S. Kwong. Bispyridoxal polyphosphates: a new class of specific intramolecular cross-linking agents for hemoglobin. *Biochem. Biophys. Res. Comm.* 156: 9-14, 1988.

Benesch, R., L. Triner, R.E. Benesch, S. Kwong, and M. Verosky. Enhanced oxygen unloading by an interdimerically cross-linked hemoglobin in an isolated perfused rabbit heart. *Proc. Natl. Acad. Sci. USA* 81: 2941-2943, 1984.

Bonaventura, J., C. Bonaventura, B. Sullivan, G. Ferruzzi, P.R. McCurdy, J. Fox, and W.F. Moo-Pen. Hemoglobin Providence. *J. Biol. Chem.* 251: 7563, 1976.

Bucci, E., A. Razynska, B. Urbaitis, and C. Fronticelli. Pseudocross-link of human hemoglobin with mono-(3,5-dibromosalicyl)-fumarate. *J. Biol. Chem.* 264: 6191-6195, 1989.

Bunn, H.F., and J.H. Jandl. Renal handling of hemoglobin. *Trans. Assoc. Am. Physicians* 81: 147-152, 1968.

Chang, T.M.S., and R. Geyer (eds.) Proceedings of II International Symposium on Blood Substitutes. *Biomater. Artif. Cells Artif. Organs*, 1988.

Chatterjee, R., E.V. Welty, R.Y. Walder, S.I. Pruitt, P.H. Rogers, A. Arnone, and J.A. Walder. Isolation and characterization of a new hemoglobin derivative cross-linked between a chains (lysine $99\alpha_1$-lysine $99\alpha_2$). *J. Biol. Chem.* 261: 9929-9937, 1986.

Chiancone, E., J.E. Norne, S. Forsen, E. Antonini, and J. Wyman. Nuclear magnetic resonance quadrupole relaxation studies of chloride binding to human oxy- and deoxyhaemoglobin. *J. Mol. Biol.* 70: 675-688, 1972.

DeVenuto, F., and A. Zegna. Blood exchange with pyridoxylated and polymerized hemoglobin solution. *Surg. Gynecol. Obstet.* 155: 342-346, 1982.

DiDonato, A., W.J. Fantl, A.S. Acharya, and J.M. Manning. Selective carboxymethylation of the α-amino groups of hemoglobin. Effect on functional properties. *J. Biol. Chem.* 258: 11890-11895, 1983.

Fantl, W.J., A. DiDonato, J.M. Manning, P.H. Rogers, and A. Arnone. Specifically carboxymethylated hemoglobin as an analogue of carbamino hemoglobin: solution and x-ray studies of carboxymethylated hemoglobin and x-ray studies of carbamino hemoglobin. *J. Biol. Chem.* 262: 12700-12713, 1987a.

Fantl, W.J., L.R. Manning, H. Ueno, A. DiDonato, and J.M. Manning. Properties of carboxymethylated, cross-linked hemoglobin A. *Biochemistry* 26: 5755-5761, 1987b.

Hess, J.R., S.O. Fadare, L.S.L. Tolentino, N.R. Bangal, and R.M. Winslow. The intravascular persistence of crosslinked human hemoglobin. In *The Red Cell Ann: Seventh Ann Arbor Conference* (G.J. Brewer, ed.) 351-360, New York: Alan R. Liss, 1989.

Groebe, D.R., M.R. Busch, T.Y.M. Tsao, F.Y. Luh, A.E. Chung, M. Gaskell,, S.A. Liebhaber, and C. Ho. High production of human α- and β-globins in insect cells. *Protein Expression and Purification* 3: 131-141, 1992.

Kluger, R., J. Wodzinska, R.T. Jones, C. Head, T.S. Fujita, and D.T. Shih. Three-point cross-linking: potential red cell substitutes from the reaction of trimesoyl tris(methyl phosphate) with hemoglobin. *Biochemistry* 31: 7551-7559, 1992.

Kumar, R. Recombinant hemoglobins as blood substitutes: a biotechnology perspective. *Proc. Soc. Exp. Biol. Med.*, in press.

Lalezari, I., P. Lalezari, C. Poyart, M. Marden, J. Kister, B. Bohn, G. Fermi, and M.F. Perutz. New effectors of human hemoglobin: structure and function. *Biochemistry* 29: 1515-1523, 1990.

Manning, J.M. Covalent inhibitor of the gelation of sickle cell hemoglobins and their effects on function. *Adv. Enzymol.* 64: 55-91, 1991.

Manning, L.R., and J.M. Manning. Influence of the ligation state and the concentration of hemoglobin A on its crosslinking by glycolaldehyde: functional properties of crosslinked carboxymethylated hemoglobin. *Biochemistry* 27: 6640-6644, 1988.

Martin de Llano, J.J., W. Jones, K. Schneider, B.T. Chait, G. Rodgers, L.J. Benjamin, B. Weksler, and J.M. Manning. Biochemical and functional properties of recombinant human sickle hemoglobin expressed in yeast. *J. Biol. Chem.* 268: 27004-27011, 1993a.

Martin de Llano, J.J., O. Schneewind, G. Stetler, and J.M. Manning. Recombinant sickle hemoglobin in yeast. *Proc. Natl. Acad. Sci. USA* 90: 918-922, 1993b.

Moss, G.S., S.A. Gould, A.L. Rosen, L.R. Sehgal, and H.L. Sehgal. Results of the first clinical trial with a polymerized hemoglobin solution. *Biomater. Artif. Cells Artif. Organs* 17: 633, 1989.

Nagai, K., and H.C. Thogersen. Generation of β-globin by sequence-specific proteolysis of a hybrid protein produced in *Escherichia coli. Nature* 309: 810-812, 1984.

Nigen, A.M., J.M. Manning, and J.O. Alben. Oxygen-linked binding sites for inorganic anions to hemoglobin. *J. Biol. Chem.* 255: 5525, 1980.

Perrella, M., L. Rossi-Bernardi, and F.J.W. Roughton. The carbamate equilibrium between CO_2 and bovine haemoglobin at 25°C. In *Oxygen Affinity of Hemoglobin and Red Cell Acid Base Status* (P. Astrup and M. Rorth, eds.), Alfred Benzon, Symposium IV, 1972.

Perutz, M.F., D. T.-b. Shih and D. Williamson. The chloride effect in human haemoglobin. A new kind of allosteric mechanism. *J. Mol. Biol.* 239: 555-560, 1994.

Shen, T.-J., N.T. Ho, V. Simplaceanu, M. Zou, B.N. Green, M.F. Tam, and C. Ho. Production of unmodified human adult hemoglobin in *Escherichia coli. Proc. Natl. Acad. Sci. USA* 90: 8108-8112, 1993.

Tam, S.C., J. Blumenstein, and J.T. Wong. Dextran hemoglobin. *Proc. Natl. Acad. Sci. USA* 73: 2118-2121, 1976.

Ueno, H., and J.M. Manning. The functional, oxygen-linked chloride binding sites of hemoglobin are contiguous within a channel in the central cavity. *J. Prot. Chem.* 11: 177-185, 1992.

Ueno, H., A.M. Popowicz, and J.M. Manning. Random chemical modification of the oxygen-linked chloride binding sites of hemoglobin: those in the central dyad axis may influence the transition between deoxy- and oxyhemoglobin. *J. Prot. Chem.* 12: 561-570, 1994.

Vandegriff, K.D., F. Medina, M. Marini, and R.M. Winslow. Equilibrium oxygen binding to human hemoglobin crosslinked between the α chains by bis(3,5-dibromosalicyl)fumarate. *J. Biol. Chem.* 264: 17824-17833, 1989.

Wagenbach, M., K. O'Rourke, L. Vitez, A. Wieczorek, S. Hoffman, S. Durfee, J. Tedesco, and G. Stetler. Synthesis of wild type and mutant human hemoglobins in *Saccharomyces cerevisiae. Bio/Technology* 9: 57-61, 1991.

Winslow, R.M. *Hemoglobin-based red cell substitutes.* Baltimore: Johns Hopkins University Press, 1992.

Yanase, H., L.R. Manning, K.D. Vandegriff, R.M. Winslow, and J.M. Manning. A recombinant human hemoglobin with asparagine-102(β) substituted by alanine has a limiting low oxygen affinity reduced marginally by chloride. *Prot. Sci.*, in press.

Chapter 7

Encapsulation of Hemoglobin in Liposomes

Alan S. Rudolph, Ph.D.

Center for Biomolecular Science and Engineering, Naval Research Laboratory, Code 6900, Washington, DC 20375-5348

ABSTRACT

One strategy to deliver hemoglobin in an oxygen carrying fluid is based on the sequestration of hemoglobin in biocompatible carriers. Much of the current work in this field is focused on the use of liposomes. Liposomes are biodegradable capsules which permit the diffusion of oxygen while encapsulating hemoglobin within an aqueous environment. The encapsulation of hemoglobin in liposomes extends the circulation persistence and alters the biodistribution of free or chemically modified hemoglobin. Numerous animal studies have shown effective oxygen delivery by liposomes with hemoglobin in models of hypovolemia and hemorrhagic shock. The focus of much work to define consequences of encapsulated hemoglobin administration has been on the effects of large-dose liposome application. This work has been directed at discerning the effects of liposome-encapsulated hemoglobin on organs of the reticuloendothelial system, particularly the liver and spleen as these are the principal sites of liposome accumulation. A number of transient effects are observed following the administration of clinically relevant doses of liposome-encapsulated hemoglobin such as a rise in liver enzymes, thrombocytopenia, leukocytosis, and complement activation. More recent studies have examined differences in the vasoactivity of encapsulated vs. free hemoglobin. These studies indicate that encapsulation markedly attenuates the vasoactivity of hemoglobins in an isolated aortic ring model. Large-scale manufacturing methods to produce sterile filtered preparations have been developed which increase the opportunities for GMP production and commercial development. This review will touch upon these issues with reference to the historic and recent literature on the safety and efficacy of liposome-encapsulated hemoglobin.

Blood Substitutes: Physiological Basis of Efficacy
Winslow et al., Editors
© Birkhäuser Boston 1995

7.1 LEH Composition and Methods of Formation

Liposomes are comprised of self-assembling amphiphilic lipids (see Figure 7.1). They form spontaneously in high-dielectric solvents like water based on the high entropic cost of exposing the hydrophobic chains of the lipids to water (Tanford 1980). Liposomes have been fabricated with a wide variety of compositions based on the desired application. Liposomes used as carriers of hemoglobin are usually comprised of double- chain phospholipids and sterols (see Table 7.1). The presence of sterols such as cholesterol results in an attenuation of thermotropic phospholipid phase transitions at which the bilayer is transiently destabilized, causing membrane fusion and leakage of encapsulated contents (Oldfield and Chapman 1971, Gregoriadis and Davis 1979, Davis and Chapman 1986). In addition, the cholesterol increases the elasticity of the membrane, which results in greater mechanical stability under shear conditions (Needham, McIntosh and Evans 1988). The

LIPOSOME-ENCAPSULATED HEMOGLOBIN

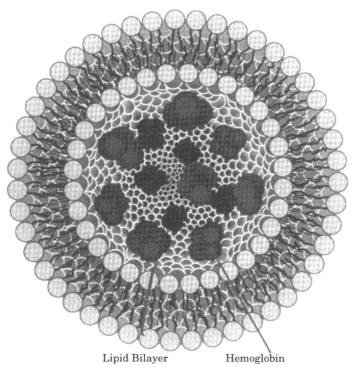

Lipid Bilayer   Hemoglobin

Figure 7.1 Diagram of liposome-encapsulated hemoglobin.

phospholipids used in various preparations have consisted mainly of egg, vegetable (soy), or synthetic phosphatidylcholines with 14-18 chain-length fatty acids (*i.e.*, dimyristoyl, dipalmitoyl, or distearoyl phosphatidylcholine). The choice of phospholipid for a liposome application can be based on production methods (see Table 7.1). For example, encapsulation efficiency increases if the liposomes are processed above their phase transition temperature, which is largely determined by the phospholipid type (Beissinger, Farmer and Gossage 1986, Bally *et al.* 1992).

The composition of lipids used can also determine the biological responses to liposomes once administered. Opsonization of serum proteins to the liposome surface initiates the recognition and phagocytic events associated with liposome removal from circulation (Patel 1992). The longer chain phospholipids have longer circulation half-times; the mechanism of this effect is not well understood (Beissinger, Farmer and Gossage 1986, Gabizon and Papahadjopoulus 1988). Thus, a large body of data on liposome-encapsulated hemoglobin has been generated with dipalmitoyl or distearoyl phosphatidylcholine. Charged phospholipids can play a large role in improved encapsulation efficiency by increasing the charge repulsion of particles and improving dispersion of liposomes during processing (Beissinger, Farmer and Gossage 1986, Bally *et al.* 1992). Charge can also play a role in biological responses to liposomes (Juliano and Stamp 1975). Most liposome-encapsulated hemoglobin preparations contain small mole percentages of negatively charged phospholipids such as dimyristoyl phosphatidylglycerol.

One proposed advantage to using liposomes to carry hemoglobin has been to extend the circulation persistence of hemoglobin by modifying the liposome surface. While liposomes demonstrated longer circulation persistence than cell free hemoglobin, there have been attempts to extend the circulation persistence even further through surface modification. Liposome formulations used for the delivery of other pharmaceuticals have employed ganglioside GM1, a sialic acid containing glycolipid or polyethylene glycol derivatives of phosphatidylethanolamine to extend the circulation persistence and increase the therapeutic value of the encapsulated agent (Lasic *et al.* 1991, Liu, Mori and Huang 1992, Parr, Bally and Cullis 1993). Investigations of extended circulation persistence with liposome-encapsulated hemoglobin formulations that have employed these agents have not resulted in significant increases in circulation persistence of liposome-encapsulated hemoglobins (Sherwood *et al.* 1993, Goins, Ligler and Rudolph 1994). This may indicate that extending circulation persistence of encapsulated hemoglobin may be difficult due to the large dose of liposomes administered in a red cell substitute application compared to other liposome indications. Surface modification of encapsulated hemoglobin may induce other advantageous properties, however, as liposome aggregation in plasma has been shown to be reduced in liposome-encapsulated hemoglobin formulations which employ polyethylene glycol surface modification (Yoshioka 1991). Additional surface modification has been examined with the use of carboxymethylchitin to shield the outer surface of the liposome (Nishiya and Chang 1993).

Liposomes self-assemble in aqueous solution into multilamellar vesicles which trap a small volume of a dissolved solute per mole of lipid compared to the total volume of solution. Processing of liposome preparations to increase the trapped volume is useful for increasing the total encapsulated hemoglobin concentration. Liposome-encapsulated hemoglobin has been fabricated with a number of traditional liposome preparation methods, including hydrodynamic shear (using a microfluidiser), detergent dialysis, sonication, mechanical agitation, and rotary evaporation (Djordjevich and Miller 1980, Farmer and Gaber 1987, Beissinger, Farmer and Gossage 1986, Jopski *et al.* 1989). To attain a useful trapped hemoglobin content, the starting concentration of hemoglobin usually is high (15-35 g/dl), which sometimes requires the stabilization of concentrated hemoglobin by conversion to deoxy or CO hemoglobin (Tsuchida 1993, Usaba *et al.* 1993). The typical total hemoglobin concentration of liposome-encapsulated hemoglobin ranges from 3-10 g/dl, depending on the starting hemoglobin concentration. It should be noted that the intraliposomal concentration is much higher (10-30%), depending on the volume percent of liposomes (or trapped volume) in a given preparation. The 20-30% encapsulation efficiency in the preparations means that cost-efficient production of LEH will require recycling of unentrapped hemoglobin, which has been demonstrated.

Additional components of liposome-encapsulated hemoglobin preparations have included stabilizers to retard the oxidation of hemoglobin and enable lyophilization (Rudolph 1988, Stratton *et al.* 1988). One of the drawbacks of hemoglobin encapsulation is that the hemoglobin is not accessible to antioxidants present in blood. Thus, the functional half-life may be different than the physical removal of particles (often expressed as circulation persistence). Hemoglobin also has been shown to interact with the lipid bilayer and result in autoxidation of lipids, generating in some cases new toxic species of lipids (Szebeni *et al.* 1985, Szebeni and Toth 1986, Itabe, Kobayashi and Inoue 1988). Membrane bound antioxidants such as alpha-tacopherol and co-encapsulated glutathione have been added to reduce methemoglobin formation and reduce interactions between hemoglobin and the lipid bilayer (Stratton *et al.* 1988). The inclusion of disaccharides in the preparation has resulted in the ability to lyophilize and reconstitute functionally and structurally intact liposome-encapsulated hemoglobin. The protective disaccharides have been shown to prevent fusion and leakage of hemoglobin upon rehydration (Rudolph 1988, Rudolph and Cliff 1990, Wang *et al.* 1992). The various preparation methods and liposome compositions used to fabricate liposome-encapsulated hemoglobin are reviewed in Table 7.1.

One important feature of liposome-encapsulated hemoglobin preparations is particle size, as it has consequences for both large-scale manufacturing methods and biological responses. The submicron particle size of most preparations has often been speculated to allow the passage of particles into vasoconstricted areas, although this has never been demonstrated. Another consideration for particle-size production characteristics of liposome-encapsulated hemoglobin is the ability to sterile filter the preparation, which requires liposome diameters that can pass through 0.45 or 0.22 micron filters. The size distribution of liposome-

Table 7.1 Liposome formulations and methods of formation for liposome-encapsulated hemoglobin.

Composition	Method of Formation	Reference
nylon	polymerization	Chang 1964
veg PC:DCP DPPC:PA:chol	agitation, sonication	Djordjevich and Miller 1980
egg PC:DPPA:chol: vitE	agitation, rotary evap.	Hunt *et al.* 1985
DCP:DMPG:chol	extrusion	Farmer and Gaber 1987
egg PC:PI:chol egg PC:PA:chol	agitation sonication	Szebeni *et al.* 1985
HSPC:DMPG:chol: vitE	microfluidization	Beissinger *et al.* 1986
DSPC:DMPG:chol: vitE	microfluidization	Rudolph *et al.* 1991
Egg PC:DPPA:chol vitE	detergent dialysis	Jopski *et al.* 1989
HSPC:myristic acid: vitE:PEG-PE	mechanical agitation	Yoshoika 1991, Sherwood *et al.* 1993
DSPC:GM1:chol: vitE	microfluidization	Goins, Ligler and Rudolph 1994

abbreviations used: PC = phosphatidylcholine, DPPC = dipalmitoyl phosphatidylcholine, BPS = bovine brain phosphatidylserine, chol = cholesterol, vitE = vitamin E, DCP = dicetylphosphate, DMPG = dimyristoyl phosphatidylglycerol, PI = phosphatidylinositol, PA = phosphatidic acid, HSPC = hydrogenated soy phosphatidylcholine, DSPC = distearoyl phosphatidylcholine, DPPA = dipalmitoyl phosphatidic acid, PEG-PE = polyethylene glycol derivative of phosphatidylethanolamine, GM1 = monosialoganglioside.

encapsulated hemoglobin that has been processed by a number of methods is often bimodal (main peak at 0.3 microns with 10-20% population at 0.8-1.0 microns). The larger-size population may be aggregates of liposomes, mediated by the adherent hemoglobin to the outer leaflet of the liposome bilayer (Rabinovici, Rudolph and Feuerstein 1990b). The adherent hemoglobin may play an important role in other manufacturing issues as well as biological effects. It was recently shown that liposome-encapsulated hemoglobin binds endotoxin and that the bound endotoxin is biologically active (Cliff, Kwasiborski and Rudolph 1994). The binding of LPS to hemoglobin has been recently examined and may contribute to the synergistic action of hemoglobin and LPS in exacerbating the consequences of sepsis (White *et al.* 1986, Roth 1994). The binding of LPS to liposome-encapsulated hemoglobin also has important implication for the quality control testing of and possible application of liposome-encapsulated hemoglobin in trauma patients with sepsis. The typical characteristics for LEH preparations made to date are reviewed in Table 7.2.

Table 7.2 Typical parameters of liposome-encapsulated hemoglobin manufactured to date.

Composition	see Table 7.1
Methemoglobin	2-15%[1]
Size	0.2 - >1.0 microns[2]
Intravesicular hemoglobin	10-30 g/dl
Total hemoglobin	3-10 g/dl[3]
P50	20-30 torr[4]
Liposome volume fraction	30-45%
Lipid concentration	100 mM
Endotoxin	0.4 Eu/ml - unknown[5]

[1]The methemoglobin value is largely determined by the hemoglobin before processing. Most liposome processing methods result in minimal increase in methemoglobin (2-5%).

[2]Liposome size is dependent on the processing methods.

[3]Total hemoglobin relates to the trapped volume of the liposome solution, which is typically 30-40%.

[4]Oxygen carrying capacity is unchanged by liposome preparation and depends on the type of hemoglobin and buffers used in processing.

[5]Endotoxin testing of hemoglobin solutions requires strict validation with both LAL and rabbit pyrogen testing due to endotoxin binding, see Cliff, Kwasiborski and Rudolph 1994.

7.2 Biological Responses to Liposome-Encapsulated Hemoglobin

Most of the biological responses to the administration of liposome-encapsulated hemoglobin are dictated by the liposome surface. Recruitment of serum proteins to the surface mediate recognition and removal of liposomes from the circulation (Patel 1992). The liposome formulation used in the fabrication of liposomes with hemoglobin results in a vehicle that is quite stable *in vivo*. There is no significant leakage of hemoglobin from the liposomes following administration, as evidenced by quantitative biodistribution studies with radiolabeled liposome-encapsulated hemoglobin (Rudolph *et al.* 1991, Phillips *et al.* 1992). These studies show no accumulation of the encapsulated tetrameric hemoglobin in the kidney, the traditional site of tetrameric hemoglobin clearance. Thus, many of the observed effects are related to the large-dose application of liposomes. The mechanism by which liposome-encapsulated hemoglobin is cleared is not yet elucidated, but new information on the role of complement may be informative (Szebeni *et al.* 1994). A summary of all of the *in vivo* studies performed with liposome-encapsulated hemoglobin can be found in Table 7.3.

7.2.1 Top Loading in Normovolemic Animals.

Administration of 10-25% blood volume with liposome-encapsulated hemoglobin (approximately 2g Hb/kg and 1g lipid/kg animal) in normovolemic animals results in a number of transient effects. A moderate increase in liver transaminases (alanine aminotransferase and aspartate transaminase) is observed following administration of 10-25% (blood volume) liposome-encapsulated hemoglobin, which returns to baseline after 24 hours (Hunt *et al.* 1985, Rudolph *et al.* 1994a). Leukocytosis is also observed over the same time course, which has been correlated to an increase in polymorphonuclear cells and not lymphocytes (Rabinovici *et al.* 1989, Rudolph *et al.* 1994a). This suggests that LEH may alter the margination of these cells. Transient thrombocytopenia also has been observed at this dose level in both rats and rabbits, which return to baseline at 1 hour postadministration (Rabinovici, Rudolph and Feuerstein 1989, Phillips *et al.* 1994). Recent studies have demonstrated that the thrombocytopenia is due to a transient sequestration of platelets in the lung (Phillips *et al.* 1994). Thrombocytopenia appears to be related to the application of liposomes and has been observed in other liposome preparations in small animals (Loughrey *et al.* 1990). The mechanism of this effect is related to the activation of complement (Loughrey *et al.* 1990). Recent studies with liposome-encapsulated hemoglobin in the rat also demonstrate that the thrombocytopenic reaction may be related to the activation of complement (Szebeni *et al.* 1994). The significance of complement activation and whether this is relevant to human liposome application is not clear as the CR1 receptor is not found on human platelets. Some of these effects were alleviated by the substitution of synthetic phosphatidylcholines for the soy-based phosphatidylcholine which had measurable lysolecithin contamination (Rabinovici, Rudolph and Feuerstein 1990b). An antagonist for platelet activating factor (BN 50739) also alleviated many of these untoward effects (Rabinovici *et al.* 1990a).

Table 7.3 Significant observations following *in vivo* administration of liposome-encapsulated hemoglobin.

Model	Animal	Major Findings	Reference
Total isovolemic exchange transfusion[1]	rat	Improved survival, maintenance of systemic PO_2 and MAP.	Djordjevich and Miller 1980 Hunt *et al.* 1985 Farmer *et al.* 1988 Ligler *et al.* 1989
50% and 70% hypovolemic hemorrhagic shock	rat, dog	When combined with hypertonic saline, increased O_2 tissue delivery, MAP and survival.	Rabinovici *et al.* 1993 Usaba *et al.* 1993
Partial exchange transfusion (25% and 50%)	rat, dog	Mild transient increase in SGOT, SGPT, BUN, bilirubin, splenomegaly and Kuppfer cell accumulation, variability in cardiac output.	Hunt *et al.* 1985 Djordjevich *et al.* 1987 Usaba *et al.* 1993 Rabinovici *et al.* 1994 Goins *et al.* 1994
Top load in normovolemic animals (10-25%)	rat, rabbit, monkey[3]	Transient increase in SGOT, SGPT, bilirubin, thrombocytopenia, leukocytosis, tachycardia, thromboxane, complement activation, hypertension, reduced cardiac output, 4-7 day transit through liver and spleen, long-term survival, no effect on hemodynamics and lung injury associated with endotoxin challenge.[2]	Rudolph *et al.* 1991 Rabinovici *et al.* 1989 Rabinovici *et al.* 1990a Phillips *et al.* 1992 Szebeni *et al.* 1994 Spirig *et al.* 1993 Sherwood *et al.* 1993

[1]Animals were bled to <6% hematocrit.

[2]Some transient effects were eliminated by using synthetic lecithin for soy-based lecithin which contained lysolecithin contamination (Rabinovici *et al.* 1990b).

[3]Monkey studies will be reported in a manuscript in preparation. Only transient leukocytosis was observed.

7.2.2 Exchange Transfusion and Hemorrhagic Shock

The first demonstrations of efficacy in total isovolemic exchange transfusion clearly demonstrated the ability of liposome-encapsulated hemoglobin to transport oxygen *in vivo* and support life below a lethal hematocrit of 2% (Djordjevich and Miller 1980, Ligler, Stratton and Rudolph 1989). The first direct demonstration of oxygen delivery by liposome-encapsulated hemoglobin was recently accomplished in a rat model of hypovolemic hemorrhagic shock (Rabinovici *et al.* 1993). Lyophilized liposome-encapsulated hemoglobin reconstituted with hypertonic saline was administered following a 70% reduction in blood volume, and peripheral tissue oxygen content was measured with an oxygen electrode in the abdominal muscle tissue. This group showed significantly higher PO_2, greater recovery of mean arterial pressure and survival than the control groups, which included hypertonic saline, lactated Ringers, and liposome-encapsulated hemoglobin in normal saline. It is interesting to note that administration of liposome-encapsulated hemoglobin in normal saline did not increase peripheral tissue oxygen tension, suggesting that this preparation of liposome-encapsulated hemoglobin did not pass peripherally vasoconstricted areas without the addition of a plasma expander.

Hemodynamic studies in partial exchange transfusion with liposome-encapsulated hemoglobin has documented recovery of mean arterial pressures in contrast to injection of a saline control which showed hypotension over the same 3-hour observation period (Rabinovici *et al.* 1992, Goins *et al.* 1994). The maintenance of mean arterial pressure was a result of opposing effects of increased total peripheral resistance and decreased cardiac output. The only previous hemodynamic study of exchange transfusion with liposome-encapsulated hemoglobin also showed maintenance of mean arterial pressure (Djordjevich *et al.* 1987). This was ascribed, however, to an increased cardiac output and vasodilation. In these limited studies, there has been no observation of hypertension, which has been observed with other hemoglobin-based blood substitutes.

7.2.3 Effect of Liposome-encapsulated Hemoglobin on Reticuloendothelial System Function

Histopathological studies following the administration of liposome-encapsulated hemoglobin show significant accumulation of material in the liver and spleen (Hunt *et al.* 1985, Rudolph *et al.* 1991, Rudolph *et al.* 1994a). The transit through these organs, following clinically relevant administration of liposome-encapsulated hemoglobin, is 4-7 days with full recovery of organ structure. Kupffer cells in the liver are principally involved with clearance of liposome-encapsulated hemoglobin. The spleen shows significant loss of red pulp with considerable filtering at early time points (Rudolph *et al.* 1994b).

Recent studies have focused on the consequences of accumulation of liposome-encapsulated hemoglobin in the reticuloendothelial system (RES). This is particularly relevant for application of blood substitutes in combat medicine, as septic challenge often accompanies blood loss. The functional status of the RES has been addressed by examination of carbon clearance following liposome-encapsulated hemoglobin administration (Beach *et al.* 1992). Carbon clearance was not significantly altered by liposome-encapsulated hemoglobin administration 3 hours following administration of 10% top load in rats and in an isolated perfused rat liver model. More recent studies have focused on the production of inflammatory cytokines in response to liposome-encapsulated hemoglobin or secondary challenge with endotoxin. Liposome-encapsulated hemoglobin does not illicit tumor necrosis factor by the RES but does retard the ability of macrophages to express tumor necrosis factor in response to an endotoxin challenge (Langdale *et al.* 1992, Rudolph *et al.* 1994a,d). The effect of liposome-encapsulated hemoglobin on the RES is clearly an area that will require additional studies. *In vivo* studies to examine secondary infectious challenge are required to determine the level of RES function following administration of a clinically relevant dose of liposome-encapsulated hemoglobin. Preliminary evidence is not conclusive (Sherwood *et al.* 1993, Spirig *et al.* 1993).

7.3 Future Research and Development

The progress in understanding the safety and efficacy of liposome-encapsulated hemoglobin has been substantial in the last decade. There are, however, areas of research that have been largely unexplored and are important to the goal of further development and eventual human testing. Recent efforts have been undertaken to understand the vasoactivity of encapsulated hemoglobin, as hemoglobin has been shown to induce significant vasoconstriction. Preliminary efforts have shown that encapsulation of hemoglobin dramatically attenuates the vasoconstrictor activity of hemoglobin (Rudolph *et al.* 1994d). Further studies will examine the interaction of liposome-encapsulated hemoglobin with the endothelium and nitric oxide binding. Additional information of effects of liposome-encapsulated hemoglobin on the RES will also be important to further development. These efforts will examine antibody production, *in vivo* sepsis models, and macrophage function following administration.

Although large-scale manufacturing methods have been developed for production of liposome-encapsulated hemoglobin (including sterile filtration methods), quality control assays are required. The interaction of the lipid bilayer and hemoglobin, as well as changes in the encapsulated hemoglobin are important components of this activity. These efforts might lead to new generations of liposome preparations and contribute to the clinical application of encapsulated hemoglobin. In addition, they will allow for studies in larger animal models, which will add to our understanding of the biological effects of this hemoglobin-based blood substitute.

7.4 References

Bally, M.B., L.D. Mayer, M.J. Hope, and R. Nayar. Pharmoacodynamics of liposomal drug carriers: methodological considerations. In *Liposome Technology, Vol. II.* (G. Gregoriadus, ed.), Boca Raton: CRC Press, 1992, pp. 27-46.

Beach, M.C., J. Morley, L. Spiryda, and S.B. Weinstock. Effects of liposome-encapsulated hemoglobin on the reticuloendothelial system. *Biomat. Artif. Cells Immobil. Biotech.* 20: 771-776, 1992.

Beissinger, R.L., M.C. Farmer, and J.L. Gossage. Liposome-encapsulated hemoglobin as a red cell surrogate. *Trans. Am. Soc. Artif. Intern. Organs* 32: 58-63, 1986.

Chang, T.M.S. Semi-permeable microcapsules. *Science* 146: 524-525, 1964.

Cliff, R.O., V. Kwasiborski, and A.S. Rudolph. A comparative study of the accurate measurement of endotoxin in liposome-encapsulated hemoglobin. *Artif. Cells, Blood Substitutes, Immobil. Biotech.*, in press, 1994.

Davis, C. and Chapman. The effect of lipid composition on the stability of liposomes *in vivo. Biochem. Soc. Trans.* 7: 680-682, 1986.

Djordjevich, L., J. Mayoral, I.F. Miller, and A.D. Ivankovich. Transfusion with synthetic erythrocytes: ability to maintain oxygen transport; histology of organs. *Crit. Care Med.* 15: 318-325, 1987.

Djordjevich, L. and I.F. Miller. Synthetic erythrocytes from lipid-encapsulated hemoglobin. *Exp. Hematol.* 8: 584-592, 1980.

Farmer, M.C. and B.P. Gaber. Liposome-encapsulated hemoglobin as an artificial oxygen carrying system. *Meth. Enzymol.* 149: 184-200, 1987.

Farmer, M.C., A.S. Rudolph, K.D. Vandegriff, M.D. Hayre, S.A. Bayne, and S.A. Johnson. Liposome-encapsulated hemoglobin: Oxygen binding and respiratory function. *Biomat. Artif. Cells Artif. Organs* 16: 289-301, 1988.

Gabizon, A. and D. Papahadjpoulos. Liposome formulations with prolonged circulation time in blood and enhanced uptake by tumors. *Proc. Natl. Acad. Sci. USA* 85: 6949-6953, 1988.

Goins, B., R. Klipper, J. Sanders, A.S. Rudolph, and W.T. Phillips. Circulation profile of [99m]technetium-labeled lipsome-encapsulated hemoglobin in a 10% or 50% rat hypovolemic shock model. *Artif. Cells, Blood Substitutes, Immobil. Biotech.* 22: 909-915, 1994.

Goins, B., F.S. Ligler, and A.D. Rudolph. Inclusion of ganglioside GM1 into liposome-encapsulated hemoglobin does not extend circulation persistence at clinically relevant doses. *Artif. Cells, Blood Substitutes, Immobil. Biotech.* 22: 9-25, 1994.

Gregoriadis, G., and C. Davis. Stability of liposomes *in vivo* and *in vitro* is promoted by their choloesterol content and the presence of blood cells. *Biochim. Biophys. Res. Comm.* 89: 1287-1293, 1979.

Hunt, C.A., R.R. Burnette, R.D. MacGregor, A. Strubbe, D.T. Lau, N. Taylor, and H. Kawada. Synthesis and evaluation of a prototypal artificial red cell. *Science* 230: 1165-1168, 1985.

Itabe, H., T. Kobayashi, and K. Inoue. Generation of toxic phospholipids during oxyhemoglobin-induced peroxidation of phosphatidylcholines. *Biochim. Biophys. Acta* 961: 13-21, 1988.

Jopski, B., V. Pirkl, H.W. Jaroni, R. Schubert, and K.H. Schmidt. Preparation of hemoglobin-containing liposomes using octylglucoside and octyltetraoxyethylene. *Biochim. Biophys. Acta* 978: 79-84, 1989.

Juliano, R.L., and D. Stamp. The effect of particle size and charge on the clearance rates of liposomes and liposome-encapsulated drugs. *Biochem. Biophys. Res. Comm.* 63: 651-658, 1975.

Langdale, L.A., R.V. Maier, L. Wilson, T.H. Pohlman, J.G. Williams, and C.L. Rice. Liposome-encapsulated hemoglobin inhibits tumor necrosis factor release from rabbit alveolar macrophages by a post-transcriptional mechanism. *J. Leuk. Biol.* 52: 679-686, 1992.

Lasic, D.D., F.J. Martin, A. Gabizon, S.K. Huang, and D. Papahadjopoulos. Sterically stabilized liposomes: a hypothesis on the molecular origin of the extended circulation times. *Biochem. Biophys. Acta* 1070: 187-192, 1991.

Ligler, F.S., L.P. Stratton, and A.S. Rudolph. Liposome-encapsulated hemoglobin: stabilization, encapsulation and storage. In: *The Red Cell: Seventh Ann Arbor Conference,* New York: Alan R. Liss, Inc., 1989, pp. 435-455.

Liu, D., A. Mori, and L. Huang. Role of liposome size and RES blockade in controlling biodistribution and tumor uptake of GM1-containing liposomes. *Biochim. Biophys. Acta* 1104: 95-101, 1992.

Loughrey, H.C., M.B. Bally, L.W. Reinish, and P.R. Cullis. The binding of phosphatidylglycerol liposomes to rat platelets is mediated by complement. *Thrombosis and Haemostasis* 64: 172-176, 1990.

Needham, D., T.J. McIntosh, and E. Evans. Thermomechanical and transition properties of dimyristoylphosphatidylcholine/cholesterol bilayers. *Biochemistry* 27: 4668-763, 1988.

Nishiya, T. and T.M.S. Chang. Toxicity of liposomes containing low mole percent of dienoyl phosphocholine to blood: use of carboxymethyl chitin to reduce toxicity. *Proceedings of the Vth International Society on Blood Substitutes*, San Diego, CA (Abstract: H118), 1993.

Oldfield, E., and D. Chapman. Effects of cholesterol and cholesterol derivatives on hydrocarbon chain mobility in lipids. *Biochem. Biophys. Res. Comm.* 43: 610-6, 1971.

Parr, M.J., M.B. Bally, and P.R. Cullis. The presence of GM1 in liposomes with entrapped doxorubicin does not prevent RES blockage. *Biochim. Biophys. Acta* 1168: 249-252, 1993.

Patel, H.M. Serum opsonins and liposomes: their interaction and opsonophagocytosis. *Critical Reviews in Therapeutic Drug Carrier Systems.* 9: 39-90, 1992.

Phillips, W.T., A.S. Rudolph, B. Goins, and R. Klipper. Biodistribution studies of liposome-encapsulated hemoglobin studied with a newly developed 99m-technetium liposome label. *Biomat. Artif. Cells Immobil. Biotech.* 20: 757-760, 1992.

Phillips, W.T., R. Klipper, A.S. Rudolph, and B. Goins. Investigation of liposome-encapsulated hemoglobin/platelet interactions using indium111-labeled platelets. *Artif. Cells, Blood Substitutes, Immobil. Biotech.* 22: 144a, 1994.

Rabinovici, R., A.S. Rudolph, and G. Feuerstein. Characterization of hemodynamic, hematological, and biochemical responses to administration of liposome-encapsulated hemoglobin in the conscious, freely moving rat. *Circ. Shock* 29: 115-132, 1989.

Rabinovici, R., A.S. Rudolph, T.L. Yue, and G. Feuerstein. Biological responses to liposome-encapsulated hemoglobin are improved by a PAF antagonist. *Circ. Shock* 31: 431-445, 1990a.

Rabinovici, R., A.S. Rudolph, and G. Feuerstein. Improved biological properties of synthetic distearoyl phosphatidyl choline-based liposome in the conscious rat. *Circ. Shock* 30: 207-219, 1990b.

Rabinovici, R., A.S. Rudolph, F.S. Ligler, E.F. Smith, and G. Feuerstein. Biological responses to exchange transfusion with liposome-encapsulated hemoglobin. *Circ. Shock* 37: 124-133, 1992.

Rabinovici, R., A.S. Rudolph, J. Vernick, and G. Feuerstein. A new salutary resuscitative fluid: Liposome-encapsulated hemoglobin/hypertonic saline solution. *J. Trauma* 35: 121-127, 1993.

Rabinovici, R., A.S. Rudolph, J. Vernick, and G. Feuerstein. Lyophilized liposome encapsulated hemoglobin: Evaluation of hemodynamic, biochemical, and hematologic responses. *Crit. Care Med.* 22: 480-485, 1994.

Roth, R.I. Hemoglobin enhances the production of tissue factor by endothelial cells in response to bacterial endotoxin. *Blood* 83: 2860-2865, 1994.

Rudolph, A.S. The freeze-dried preservation of liposome-encapsulated hemoglobin: A potential blood substitute. *Cryobiology* 25: 277-284, 1988.

Rudolph, A.S., and R.O.C. Cliff. Dry storage of liposome-encapsulated hemoglobin: a blood substitute. *Cryobiology* 27: 585-590, 1990.

Rudolph, A.S., R.W. Klipper, B. Goins, and W.T. Phillips. *In vivo* biodistribution of a radiolabeled blood substitute: ^{99m}Tc-labeled liposome-encapsulated hemoglobin in an anesthetized rabbit. *Proc. Natl. Acad. Sci. USA* 88: 10976-10980, 1991.

Rudolph, A.S., R.O. Cliff, B.J. Spargo, and H. Spielberg. Transient changes in the mononuclear phagocyte system following administration of the blood substitute, liposome-encapsulated hemoglobin. *Biomaterials* 15: 796-804 1994a.

Rudolph, A.S., R.O. Cliff, R. Klipper, B.A. Goins, and W.T. Phillips. Circulation persistence and biodistribution of lyophilized liposome-encapsulated hemoglobin: an oxygen-carrying resuscitative fluid. *Crit. Care Med.* 22: 142-150, 1994b.

Rudolph, A.S., H. Spielberg, B.J. Spargo, and N. Kossovsky. Histolopathological study following the administration of liposome-encapsulated hemoglobin in the normovolemic rat. *J. Biomed. Mat. Res.,* in press, 1994c.

Rudolph, A.S., R.O. Cliff, V. Kwasiborski, P. Heible, T. Sulpizio, and G. Feuerstein. Macrophage interactions and vasoactivity of liposome-encapsulated hemoglobin, an artificial oxygen carrying resuscitative fluid. *Intensive Care Med.,* in press, 1994d.

Sherwood, R., D. McCorMick, S. Zheng, R. Beissinger, and F. Martin. Influence of steric stabilization on immunosuppressive activity of liposome-encapsulated hemoglobin. *Proceedings of the Vth International Symposium on Blood Substitutes,* San Diego, CA, #H20, 1993.

Spirig, A., G. Feuerstein, A.S. Rudolph, P. Bugelsky, J. Richardson, D. Dietz, and R. Rabinovici. Lyophilized liposome-encapsulated hemoglobin treatment in LPS-induced sepsis. *Proceedings of the Vth International Symposium on Blood Substitutes,* San Diego, CA, #H19, 1993.

Stratton, L.P., A.S. Rudolph, W.K. Knoll, and M.C. Farmer. The reduction in methemoglobin levels by antioxidants. *Hemoglobin* 12: 353-368, 1988.

Szebeni, J., E.E. DiIorio, H. Hauser, and K.H. Winterhalter. Encapsulation of hemoglobin in phospholipid liposomes: characterization and stability. *Biochemistry* 24: 2827-2832, 1985.

Szebeni, J., and K. Toth. Lipid peroxidation in hemoglobin containing liposomes: effects of membrane lipid composition and cholesterol content. *Biochim. Biophys. Acta* 857: 139-145, 1986.

Szebeni, J., N.M. Wassef, H. Spielberg, A.S. Rudolph, and C.R. Alving. Complement activation in rats by liposome-encapsulated hemoglobin. *Biochem. Biophys. Res. Comm.* 205: 255-263, 1994.

Tanford, C. *The Hydrophobic Effect*. New York: John Wiley and Sons, 1980.

Tsuchida, E. Stabilized hemoglobin vesicles. *Proceedings of the Vth International Symposium on Blood Substitutes*, San Diego, CA, #H17, 1993.

Usaba, A., R. Motoki, K. Sakaguchi, K. Suzuki, and T. Kamitani. Effect of neo red cells on hemodynamics and blood gas transport in canine hemorrhagic shock and its safety for vital organs. *Proceedings of the Vth International Symposium on Blood Substitutes*, San Diego, CA, #H21, 1993.

Wang, L., S. Takeoka, E. Tsuchida, S. Toduyama, T. Mashiko, and T. Satoh. Preparation of dehydrated powder of hemoglobin vesicles. *Polymers for Advanced Technologies* 3: 17-21. 1992.

White, C.T., A.J. Murrat, D.J. Smith, J.R. Greene, and R.B. Bolin. Synergistic toxicity of endotoxin and hemoglobin. *J. Lab. Clin. Med.* 108: 132-137, 1986.

Yoshioka, H. Aggregation of liposome-encapsulated hemoglobin in plasma is inhibited by polyethylene glycol surface modification. *Biomaterials* 142: 861-864, 1991.

Chapter 8

Stability and Toxicity of Hemoglobin Solutions

Kim D. Vandegriff, Ph.D.

Department of Medicine, School of Medicine, University of California, San Diego, Veterans Affairs Medical Center (111-E), 3350 La Jolla Village Drive, San Diego, California 92161

8.1 Introduction

The anticipated clinical application of large amounts of highly concentrated hemoglobin as an acellular O_2 carrier has presented an array of challenges to the protein engineer. The simplest of these have been met: designing chemical and mutant modifications that reduce or eliminate tetramer-dimer dissociation and ones that covalently modulate O_2 affinity. Some of these modified proteins have been tested in humans, and while the results are promising, they point out that difficult challenges are still ahead. The principal challenge now is to reduce or to eliminate toxic side effects, because it is unlikely that a red cell substitute that is not virtually safe will compete successfully with red blood cells for clinical use.

Properties of cell-free hemoglobin that potentially may lead to *in vivo* toxicity are heme-globin dissociation and denaturation (producing insoluble globin), production of O_2-free radicals, extravasation, and inhibition of NO activity by binding to hemoglobin or through oxidation reactions. Common to at least a few of the problems is hemoglobin structural stability, which also raises logistic concerns about the long-term storage of these solutions.

Basic mechanisms of hemoglobin stability must be understood to predict and, hopefully, avoid toxic side effects. This chapter discusses the biochemistry of hemoglobin oxidation and denaturation and assumes some basic understanding of hemoglobin chemistry. It addresses issues of stability, including mechanisms of hemoglobin degradation, new strategies for designing stable molecules using site-directed mutagenesis, and the

Blood Substitutes: Physiological Basis of Efficacy
Winslow et al., Editors
© Birkhäuser Boston 1995

potential toxicities of acellular hemoglobin solutions, particularly endothelial cytotoxicity.

8.2 Hemoglobin Denaturation inside Red Blood Cells

Hemoglobin is an inherently unstable molecule. It is a heme-containing
metalloprotein that binds O_2 reversibly at four reactive sites. Because
iron is a transition metal, hemoglobin also undergoes a reversible change
in oxidation state from the ferrous (Fe^{2+}) to the ferric (Fe^{3+}) or methemoglobin form, making it an effective redox catalyst. During autoxidation,
hemoglobin becomes oxidized by molecular oxygen as it dissociates from
oxyhemoglobin, removing an electron from heme iron and creating a loss
in the protein's function and stability: O_2 does not bind to ferric hemes,
and the interaction between ferric heme and globin is weakened. Once
heme dissociates, apohemoglobin unfolds and becomes insoluble at
physiologic pH and temperature.

The long-term survival of red blood cells depends on their ability to prevent this sequence of events from hemoglobin oxidation to denaturation.
Precipitated, denatured hemoglobin molecules form Heinz bodies on the
inside surface of the red cell membrane, which can cause oxidative damage to the membrane (Jacob and Winterhalter 1970, Winterbourn 1990).
Heme is a pro-oxidant that reacts with activated oxygen species, promoting lipid peroxidation, protein oxidation and inter-protein cross-linking,
ultimately leading to hemolysis (Vincent 1989). In sickle cells, the
amount of membrane-bound heme is directly related to the degree of hemolysis (Kannan, Labotka and Low 1988).

Red blood cells contain complex methemoglobin reductase and antioxidant systems to help prevent these reactions. And in normal red blood
cells, methemoglobin makes up no more than 3% of erythrocytic hemoglobin. Methemoglobinemias, however, can occur pathologically when
the red cell reductase or antioxidant systems are impaired or when genetic mutations of hemoglobin make the molecule more susceptible to
oxidative reactions.

Mutation of even a single hemoglobin amino acid residue can adversely
affect its molecular stability, and some of these hemoglobin variants are
prone to denaturation. A loss in stability can occur through structural
changes that make the molecule more flexible, opening up the heme
pocket to water or anions and increasing its rate of oxidation. Other mutations disrupt secondary structure or alter hydrophobic or hydrophilic
regions and promote unfolding of the polypeptide chains. Two particularly vulnerable areas of the molecule are the $\alpha_1\beta_2$ interface and the
heme pocket. Some natural, unstable mutants are listed in Table 8.1,
and a more extensive list can be found in Table 13-1 in Bunn and Forget
(1986).

Table 8.1 Some natural, unstable hemoglobin mutants.

Torino	αCD1(Phe $\rightarrow$ Val)	Loss of heme steric contact
Hirosaki	αCD1(Phe $\rightarrow$ Leu)	Loss of heme steric contact
Iwata	αF8(His $\rightarrow$ Arg)	Hemichrome formation and heme loss
Setif	αG1(Asp $\rightarrow$ Tyr)	Disruption of subunit contact
St. Louis	βB10(Leu $\rightarrow$ Gln)	Introduction of interior dipole
Genova	βB10(Leu $\rightarrow$ Pro)	Disruption of α helix
Hammersmith	βCD1(Phe $\rightarrow$ Ser)	Loss of heme steric contact
Louisville	βCD1(Phe $\rightarrow$ Leu)	Loss of heme steric contact
Warsaw	βCD1(Phe $\rightarrow$ Val)	Loss of heme steric contact
Zürich	βE7(His $\rightarrow$ Arg)	Opening of distal heme pocket
Bicetre	βE7(His $\rightarrow$ Pro)	Disruption of α helix
I-Toulouse	βE10(Lys $\rightarrow$ Glu)	Loss of heme electrostatic contact
Sydney	βE11(Val $\rightarrow$ Ala)	Gap in distal heme pocket
Bristol	βE11(Val $\rightarrow$ Asp)	Introduction of interior charge
Mozhaisk	βF8(His $\rightarrow$ Arg)	Heme loss
Istanbul, St. Etienne	βF8(His $\rightarrow$ Gln)	Heme loss
Newcastle	βF8(His $\rightarrow$ Pro)	Heme loss
Köln	βFG5(Val $\rightarrow$ Met)	Disruption of subunit contact
Southamption	βG8(Leu $\rightarrow$ Pro)	Disruption of α helix
Tübingen	βG8(Leu $\rightarrow$ Gln)	Introduction of interior dipole

Adapted from Bunn and Forget (1986).

8.3 Adverse Effects of Acellular Hemoglobin

During hemolysis, free hemoglobin and its breakdown products are normally removed from the circulation by a number of plasma proteins. Haptoglobin binds hemoglobin dimers, hemopexin and albumin bind heme, and transferrin binds iron. The complexes are then transported from the intravascular space to the liver, spleen and bone marrow for metabolic processing. However, during severe hemolysis, these serum proteins can become saturated, causing the concentration of free heme in plasma to rise (Muller-Eberhard *et al.* 1968). This latter situation may parallel applications of hemoglobin-based blood substitutes in which large amounts of the acellular hemoprotein are directly exposed to plasma proteins.

When cell-free hemoglobin persists in the circulation, a number of toxic reactions occur: (i) Dissociated hemoglobin dimers not bound to haptoglobin are filtered through the kidneys, potentially injuring kidney tubules. (ii) In the absence of red cell methemoglobin reductase and other cellular antioxidants, hemoglobin oxidation takes place uninhibited, and methemoglobin, superoxide anion, and hydrogen peroxide are formed. (iii) Heme dissociates from methemoglobin, and free heme and iron catalyze reactions with superoxide and hydrogen peroxide to produce highly reactive oxygen free-radical species that add or remove electrons from biological molecules and propagate a chain reaction of free radical production.

All hemoglobin products that are being developed as acellular resuscitation fluids have altered molecular structures, either through site-specific chemical or genetic cross-linking, but because even a single mutation can markedly affect hemoglobin function and stability, each new product has to be studied independently for these effects. Cross-linking prevents hemoglobin dissociation, and so renal toxicity has become less of an issue, but the potential for oxidative catalysis of free radical reactions remains an active area of research.

8.4 Structural Interactions between Heme and Globin

In heme, iron is bound through four coordinate bonds to the pyrrole nitrogens of protoporphyrin IX. In hemoglobin and myoglobin, a fifth coordinate bond is formed between heme iron and the globin at the ε-NH group of the proximal His at the eighth position of the F helix. (O_2 binds distal to the heme as a sixth coordinate iron ligand.) Heme partitions into a pocket between the E and F helices of α and β subunits in hemoglobin, primarily through hydrophobic interactions. Selected residues in the α- and β-subunit heme pockets are shown in Figure 8.1.

The apolar heme vinyl groups react with the hydrophobic interior of the pocket, and the charged heme propionate groups point outward toward the hydrophilic surface of the protein. In human hemoglobin, the heme-7-propionates interact with Lys residues at positions E10 on both subunits. In α subunits, the heme-6-propionate interacts electrostatically with a His residue at position CD3. In β subunits, CD3 is a Ser residue that provides less overall interaction to stabilize the heme.

Oxidation of heme iron from Fe^{2+} to Fe^{3+} weakens the fifth coordinate bond to the proximal His and increases the probability of heme loss (Bunn and Jandl 1968). The pathways involved in hemoglobin denaturation are presented in Figure 8.2. The first step is oxidation of the heme iron.

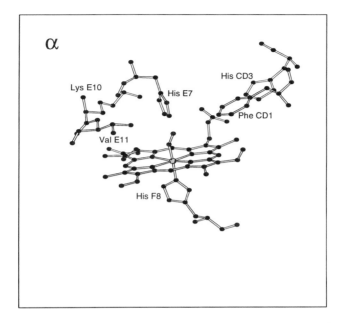

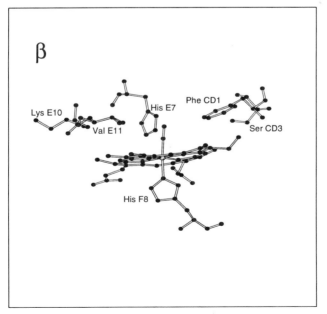

Figure 8.1 Diagram of the α- and β-subunit heme pockets in human hemoglobin. The heme ring is show with O_2 bound. (Courtesy of Lynn TenEyck, Supercomputer Center, University of California, San Diego).

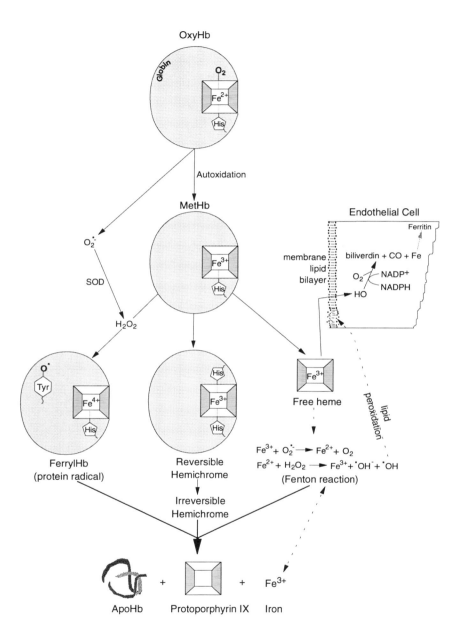

Figure 8.2 Diagram of possible pathways of hemoglobin denaturation starting with oxyhemoglobin. The proximal His is shown, and distal His is shown
bound to the ferric heme in a reversible hemichrome. SOD = superoxide dis-
mutase; HO = heme oxygenase.

8.5 Mechanisms of Hemoglobin Autoxidation

In either isolated chains or in the intact hemoglobin molecule, α subunits have a lower redox potential than β subunits (Banerjee and Cassoly 1969, Banerjee and Lhoste 1976), and as a result, α subunits oxidize more readily (Mansouri and Winterhalter 1974). Furthermore, hemoglobin dimers oxidize nearly 20 times more rapidly than intact tetramers, presumably due to the fewer number of conformational constraints within dimers (Zhang, Levy and Rifkind 1991).

There are two mechanisms of autoxidation of the hemoproteins, hemoglobin and myoglobin, that bind O_2 reversibly as O_2-carrier and storage proteins, respectively. Both mechanisms involve production of the superoxide anion ($O_2^-\cdot$), and the difference in susceptibility of the subunits to oxidation is reflected by their rates of superoxide dissociation (*i.e.*, $O_2^-\cdot$ dissociates faster from α subunits) (Demma and Salhany 1979).

The first mechanism of autoxidation involves the transfer of an electron from ferrous heme to bound O_2, producing ferric heme and the direct dissociation of superoxide (Eq. 1) (Weiss 1964):

(1) $\text{Hb}(\text{Fe}^{2+})O_2 \rightarrow \text{Hb}(\text{Fe}^{3+}) + O_2^-\cdot$

The transition of O_2 bound to ferrous heme to that of $O_2^-\cdot$ bound to low-spin ferric heme occurs when an electron in one of the iron d orbitals is transferred to an unoccupied π^* orbital of bound O_2. The electron is lost from iron as superoxide dissociates, leaving ferric heme.

The second mechanism of autoxidation involves a nucleophile-mediated exchange of an electron from ferrous iron to unbound O_2 (Wallace *et al.* 1982).

(2) $\text{HbO}_2 \leftrightarrow \text{Hb}(\text{Fe}^{2+}) + O_2$

(3) $\text{Hb}(\text{Fe}^{2+}) + N \leftrightarrow \text{Hb}(\text{Fe}^{2+})(N)$

(4) $\text{Hb}(\text{Fe}^{2+})(N) + O_2 \rightarrow \text{Hb}(\text{Fe}^{3+})N + O_2^-\cdot$

In this reaction, bound O_2 must first dissociate from the ferrous hemoprotein (Eq. 2). A nucleophile (N) then binds to the unliganded hemoprotein (Eq. 3), and oxidation of the ferrous heme iron occurs as an electron is transferred to unbound O_2 (Eq. 4). Again, superoxide is produced, and in this case, the nucleophile becomes a ferric heme ligand. It has been suggested that the distal His (E7) can act as an endogenous nucleophile that displaces $O_2^-\cdot$, creating a transient hemichrome species (Rifkind *et al.* 1994) (see below, Section 8.6.1).

The mechanism of autoxidation, either direct dissociation of the superoxide anion (Eq. 1) or nucleophilic displacement (Eqs. 2-4), depends on O_2 concentration. Using myoglobin, a single-subunit hemoprotein, Brantley *et al.* (1993) have shown that at low O_2 concentration, where dissociation of oxymyoglobin to deoxymyoglobin and O_2 is favored, the latter mechanism in Eqs. 2-4 predominates. At high O_2 concentrations, where oxy-myoglobin is favored, direct dissociation of superoxide occurs (Eq. 1). In air at 37°C, direct dissociation of superoxide is the primary mechanism, and protonation of bound O_2 is required (Brantley *et al.* 1993). Lower pH accelerates hemoprotein autoxidation because the protein-$(Fe^{2+})O_2$ complex has to be protonated before superoxide can dissociate. As proposed by Shikama (1984), an unprotonated, negatively charged superoxide anion would simply rebind to hemin.

8.5.1 Effect of O_2 Affinity

In general for myoglobin and for hemoglobin, there is a strong negative correlation between O_2 affinity and the rate of autoxidation (k_{ox}). This is true for modified hemoglobins (Macdonald *et al.* 1991) and for site-specifically mutated myoglobins (Springer *et al.* 1989, Brantley *et al.* 1993) (Figure 8.3). Therefore, hemoglobins that are designed as cell-free

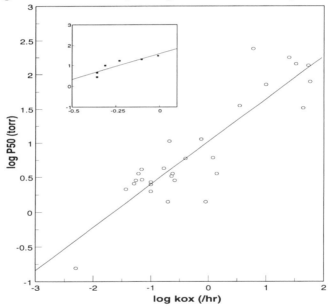

Figure 8.3 Correlation between P50 and rates of autoxidation (k_{ox}) for myoglobin and site-directed mutant myoglobins (adapted from Brantley *et al.* 1993) and (inset) for hemoglobin and chemically modified hemoglobins (adapted from Macdonald *et al.* 1991).

O_2 carriers to have low O_2 affinity will probably also have high rates of autoxidation. However, other effects can occur. For example, glutaraldehyde-polymerized hemoglobin has a decreased redox potential and a higher rate of autoxidation but also higher O_2 affinity (Guillochon, Esclade and Thomas 1986). For hemoglobin-based blood substitutes, the optimal balance between O_2 affinity and rate of autoxidation is not known. Nevertheless, it is clear that rates of oxidiation should be kept minimal to prevent hemoglobin denaturation and the production of oxidative species.

8.5.2 Effects of Distal Pocket Mutations

John S. Olson's group at Rice University and in collaboration with others have used myoglobin to study the effects of site-specific mutations on autoxidation rates. Several features have been revealed about the heme pocket. Some of their results are summarized below and in Table 8.2.

<u>Residue E7</u> (distal His): In native myoglobin (and in α subunits of human hemoglobin), distal His forms a hydrogen bond with bound O_2 that is critical for maintaining heme iron in the ferrous state by both stabilizing and inhibiting protonation of bound O_2. The loss of this hydrogen bond has dramatic effects on O_2 affinity and rates of autoxidation. Substituting distal His in myoglobin with an apolar residue increases its rate of autoxidation by 100-800 times, depending on the size of the apolar side chain (*i.e.*, the smaller the side chain, the higher the rate) (Brantley *et al.* 1993).

The smallest increase in the rate of autoxidation by a mutation of the distal His in sperm whale myoglobin in the study of Brantley *et al.* (1993) was E7(His $\rightarrow$ Gln), which showed only a 4-fold increase in rate. Interestingly, elephant myoglobin has an E7 distal Gln, and compared with human myoglobin with an E7 distal His, its rate of azide oxidation is 4.5-fold lower, even though these two myoglobins possess similar O_2 affinities (Romero-Herrera *et al.* 1981). The E7 residue is not the only difference in the heme pocket of elephant myoglobin compared with human or sperm whale myoglobins. The elephant protein also contains a Phe residue at position B10 (J.S. Olson, personal communication), which may have additional significant effects on stability (see below).

The results for the single-site E7 Gln sperm whale myoglobin mutant and for native elephant myoglobin are consistent as well with the recent report of a site-directed myoglobin double mutant E7(His $\rightarrow$ Gln)/B10 (Leu $\rightarrow$ Tyr) that has a lower rate of autoxidation but unaltered O_2 binding properties (Brunori 1994). This double mutant has the same residues at positions E7 and B10 as *Ascaris* hemoglobin in which a hydrogen bond forms between bound O_2 and these two residues (De Baere *et al.* 1994). Apparently, the additional stability of bound O_2 in the myoglobin double mutant reported by Brunori slows its rate of autoxidation.

In Hemoglobin Zürich, an unstable hemoglobin mutant, Arg is substituted for the E7 distal His in β subunits (see Table 8.1). The larger distal side chain of Arg moves out of the heme pocket to the surface of the molecule, leaving the heme pocket more exposed to solvent (Tucker *et al.* 1978).

Table 8.2. Autoxidation, hemin loss, and stability for native, wild-type, and site-directed mutant myoglobins.

	Species	k_{ox}[1]	k_{-H}[1]	Stability[1]	Reference[2]
		(hr^{-1})	(hr^{-1})		
native	sw[3]	0.06	~0.01	125,000	a,b
wild-type	sw	0.05	~0.01	70,000	a,b
native	pig	0.06	----	----	a
wild-type	pig	0.07	<0.01	600	a,b
wild-type	human	----	~0.01	2000	b
B10 (Leu → Ala)	sw	0.24	----	----	c
(Leu → Val)	sw	0.23	----	----	c
(Leu → Phe)	sw	0.005	<0.01	100,000	b,c
(Leu → Asn)	sw	----	0.10	<10	b
CD1 (Phe → Val)	sw	----	2	1400	b
(Phe → Ile)	sw	----	2	28,000	b
CD3 (Lys → Ser)	pig	0.4	----	----	a
(Lys → His)	pig	0.17	----	----	a
(Lys → Arg)	pig	0.10	----	----	a
(Lys → Glu)	pig	1.2	----	----	a
(Arg → Glu)	sw	----	0.2	----	d
E7 (His → Ala)	sw	58	0.4	330,000	a,b
(His → Leu)	sw	10	0.2	740,000	a,b
(His → Phe)	sw	6	0.08	1,600,000	a,b
(His → Gln)	sw	0.21	0.12	50,000	a,b
E10 (Thr → Ala)	pig	0.9	----	----	a
(Thr → Arg)	pig	0.04	----	----	a
E11 (Val → Ala)	sw	0.26	0.1	26,000	a,b
(Val → Phe)	sw	0.07	~0.01	500,000	a,b
(Val → Asn)	sw	----	~0.01.	200	b
(Val → Ser)	sw	----	0.1	800	b
(Val → Thr)	sw	----	<0.01	7400	b
G8 (Ile → Phe)	sw	----	~0.01.	330,000	b
(Ile → Thr)	sw	----	0.15	2900	b

[1] Conditions: k_{ox} (the rate of autoxidation) and k_{-H} (the rate of heme loss) were measured at 37°C and pH 7.0; stability was measured at 25°C. Stability refers to the overall stability of the apoglobin (see Hargrove *et al.* 1994a for a more exact definition). The higher the number for stability, the greater the stability of the apoglobin.

[2] a, Brantley *et al.* 1993; b, Hargrove *et al.* 1994a; c, Carver *et al.* 1992; d, Hargrove *et al.* 1994b

[3] sw=sperm whale

Residue E10: Pig myoglobin E10(Thr → Arg) has a slightly lower rate of autoxidation compared with the native protein. The additional positive charge in the heme pocket at a position other than distal His inhibits protonation of bound O_2 and oxidation of heme (Brantley *et al.* 1993). In α and β subunits of human hemoglobin, the E10 residues are Lys, which by this mechanism should already be unfavorable to oxidation.

Residue B10: The mutation B10(Leu → Phe) has a profound effect on both the rate of autoxidation and the O_2 affinity of myoglobin (the rate of autoxidation decreases 10-fold, and O_2 affinity increases 15-fold). It appears that by increasing the size of the B10 residue and decreasing access of the heme pocket to H_2O, bound O_2 becomes stabilized (Carver *et al.* 1992). This mutant is so stable that only ferrous myoglobin was produced in the *E. coli* expression system reported by Carver *et al.* (1992). In contrast, when the B10 residue is decreased in size, as in myoglobin-mutants B10(Leu → Ala) or B10(Leu → Val), oxidation is accelerated (Brantley *et al.* 1993).

Hemoglobin St. Louis [βB10(Leu → Gln)] is another natural, unstable hemoglobin variant (see Table 8.1). The substitution of Gln for Leu expands and increases the polarity of the distal pocket by binding a H_2O molecule between B10(Gln) and distal His, causing a high rate of oxidation of β subunits (Fermi and Perutz 1981).

8.6 Pathways of Hemoglobin Denaturation Following

Autoxidation

8.6.1 Hemichrome Formation

Hemichromes are low-spin forms of ferric hemoproteins. They form spontaneously when the ferric protein structure becomes distorted, allowing the heme iron to bind to the distal His to make a reversible hemichrome (see Figure 8.2) or, upon further distortion of the globin structure, to allow the heme iron to bind to another endogenous protein residue, making an irreversible, insoluble hemichrome. Compared with mammalian myoglobins, mammalian hemoglobins have a greater tendency to form hemichromes, perhaps because the distal pocket in hemoglobin is more flexible (Levy, Alston and Rifkind 1984). As a further note, both native and αα-cross-linked human hemoglobins show a greater propensity for hemichrome formation in the presence of bacterial endotoxins (Kaca *et al.* 1995).

8.6.2 Oxidation of Ferric Heme

Ferric heme or hemoproteins are oxidized further to a hypervalency ferryl (Fe^{4+}) state by H_2O_2, apparently through a heterolytic cleavage of the peroxide bond (Giulivi and Davies 1994). Ferryl hemoglobin is a highly reactive intermediate (similar to peroxidase series compound II) that is formed from one of the two oxidizing equivalents of H_2O_2. The second oxidizing equivalent may exist transiently as a porphyrin radical that can then react quickly with a globin residue near the heme group to create a highly reactive protein radical (see Figure 8.2). The protein radical most likely resides on an aromatic residue, and in sperm whale myoglobin, it moves from one Tyr residue to another (Wilks and Ortiz de Montellano 1992).

Myoglobin is converted to an oxidase when it is treated with even small amounts of H_2O_2. It has been suggested that the interaction between ferryl heme and the globin radical forms a modified globin-heme adduct that contains a more open active site with oxidase activity. This leads to further H_2O_2 production and finally complete protein degradation (Osawa and Korzekwa 1991). Hemoglobin, on the other hand, does not produce an oxidase when treated with H_2O_2 (Osawa et al. 1993), but alterations in the protein structure introduced by a protein radical and/or ferryl heme may be involved in hemichrome formation (Kindt et al. 1992).

Stereochemical differences between unmodified and cross-linked hemoglobins alter their ability to form or to react with free radicals. For example, there is a large distinction in the rate of formation and stability of ferryl intermediates for native, αα-cross-linked or monosubstituted β-modified human hemoglobins during enzymatic oxidation. αα-Cross-linked hemoglobin maintains a higher level of the reactive ferryl intermediate (Alayash et al. 1992). Consequently, these hemoglobins show a wide variation in susceptibility to further oxidative damage (Osawa et al. 1993). Perhaps because of its relatively stable ferryl intermediate, αα-cross-linked hemoglobin is the most susceptible to oxidative damage compared with these other hemoglobins; it is rapidly and completely destroyed after treatment with excess H_2O_2. Native hemoglobin has intermediate susceptibility, and the β-modified hemoglobin is the most resistant in this series. These effects appear to involve conformational differences that may be unrelated to O_2 affinity (Alayash et al. 1992).

In addition to self-destruction, ferryl hemoglobin and ferryl myoglobin are capable of oxidative tissue injury by promoting a chain reaction of lipid peroxidation (Dix et al. 1985, Cadenas 1989).

8.6.3 Heme Loss

The linkage between globin and heme is markedly weakened after oxidation of heme iron and some distortion in subunit structure, which may or may not be a form of hemichrome (see Figure 8.2). Estimates of the activation energies of hemin dissociation suggest that the distortion is due to local rather than global protein unfolding (Yamamoto and La Mar 1986).

Heme-globin affinity is determined primarily by the rate of hemin dissociation, and at pH 7.0 and 37°C, hemin loss from hemoglobin occurs approximately 2-3 orders of magnitude faster than from myoglobin (Hargrove *et al.* 1994b). Hemin loss from hemoglobin dimers is faster than from intact tetramers (Benesch and Kwong 1990), and α and β subunits lose their hemes at different rates (Bunn and Jandl 1968, Banerjee and Lhoste 1976). The rates of hemin loss from oxidized α and β subunits at 37°C and pH 7.0 are 0.6 and 7.8 hr^{-1}, respectively, (Hargrove *et al.* 1994b); at 20°C and pH 9.0, the comparative rates are approximately 0.2 and 2 hr^{-1} (Vandegriff and Le Tellier 1994).

Rates of hemin loss from chemically modified hemoglobins fall into at least two classes: (1) From site-specifically cross-linked hemoglobins (cross-linked between either α or β subunits), the rates are comparable to those of unmodified, native human hemoglobin (*i.e.*, ~0.2 and 2 hr^{-1} at 20°C and pH 9.0). (2) From nonspecifically, glutaraldehyde- or glycolaldehyde-polymerized hemoglobins, the rates are 10-fold higher (Vandegriff and Le Tellier 1994). These aldehyde reagents react with hemoglobin nondiscriminately, producing a heterogeneous population of cross-linked polymers. A possible explanation for their higher rates of hemin loss comes from electron paramagnetic resonance and Mössbauer spectroscopic studies of hemoglobin reacted with glutaraldehyde which show that the proximal His-Fe bond is lengthened and, therefore, weakened after polymerization (Chevalier *et al.* 1990).

The faster rate of hemin loss from β subunits is probably due to differences between the subunits near the proximal His bond or at the CD corner. A Phe residue is conserved at position CD1 in all globins and appears to be critical for heme binding (Dickerson and Geis 1983) (see Figure 8.1). Replacement of Phe with smaller residues leaves a hole in the heme pocket that destabilizes heme binding and leads to altered subunit assembly and protein denaturation (Honig *et al.* 1990). Naturally occurring hemoglobins with substitutions at this position, including hemoglobins Hammersmith βCD1(Phe → Ser), Louisville βCD1(Phe → Leu), and Warsaw βCD1(Phe → Val), are unstable mutants (see Table 8.1). *In vivo*, these mutants are heme depleted and readily form Heinz bodies (Jacob and Winterhalter 1970). In myoglobin, mutation of the CD1 residue to either a Val or Ile residue greatly accelerates hemin loss and decreases apoglobin stability (Hargrove *et al.* 1994a) (see Table 8.2).

The rates of hemin loss for a few natural hemoglobin mutants have been reported by Smith *et al.* (1984). Hemoglobins Hammersmith βCD1(Phe → Ser) and Köln βFG5(Val → Met), which are both unstable, exhibit 10-20 fold higher rates of hemin loss compared with native protein. Hemoglobins St. Mandé βG4(Asn → Tyr) and Hôtel Dieu βG1(Asp → Gly) have substitutions at the $\alpha_1\beta_2$ interface, and while they are not considered to be unstable, they still exhibit rates of hemin loss that are 3-4-fold higher than those of native human hemoglobin. Hemoglobin Malmö βFG3(His → Gln), a stable mutant, loses hemin at the same rate as native hemoglobin.

Rates of hemin loss from mutated hybrid hemoglobins in which the distal His in one type of subunit is replaced by Gly suggest that hemin loss from the native subunits is accelerated after hemin dissociates from the less stable mutant subunits (Hargrove *et al.* 1994b).

Interactions between heme propionates and globin residues are important for stabilizing bound heme. As already mentioned, Lys residues at position E10 of both α and β subunits in human hemoglobin interact electrostatically with heme-7-propionate groups. In α subunits, His at position CD3 forms an additional hydrogen bond with the heme-6-propionate. In β subunits, the CD3 residue is Ser, which appears not to interact (Shaanan 1983) (see Figure 8.1). In mammalian myoglobins, CD3 is either Lys or Arg, and in addition to interaction with the heme-6-propionate, these residues form a hydrogen bonding network between His E7, Thr E10, and coordinated water molecules in the heme pocket (Quillin *et al.* 1993). In sperm whale myoglobin, substitution of CD3 Arg with Glu increases the rate of hemin loss by 20-fold at pH 7.0 (Hargrove *et al.* 1994b).

In addition to electrostatic contacts, distal pocket residues provide important steric contacts with the porphyrin. Substitution in sperm whale myoglobin of either His E7 or Val E11 with Ala increases the rate of hemin loss by 10-40-fold at pH 7.0 (Hargrove *et al.* 1994a).

So far, only two mutant myoglobins have been prepared that have decreased rates of hemin loss at pH 7.0. These are B10(Leu → Phe) and E11(Val → Thr). E11 Thr stabilizes bound heme by forming an extra hydrogen bond with coordinated water (Smerdon *et al.* 1991). Its apoglobin is much less stable, however, and it readily unfolds once heme dissociates. The B10 Phe apoglobin is nearly as stable as native myoglobin and is expressed in *E. coli* at amounts comparable to that of the wild-type protein (Hargrove *et al.* 1994a). (See Table 8.2 for a summary of results with myoglobin mutants.)

8.6.4 Protein Unfolding and Apoglobin Stability

The most complete study of the process of denaturation of a hemoprotein has been made using myoglobin as a prototype (Hughson, Wright and Baldwin 1990, Hargrove *et al.* 1994a). Myoglobin has a folding pattern of helical segments (A-H) similar to that of hemoglobin subunits, with only minor differences (*e.g.*, the 5-residue D helix is absent in α chains). The same amino acid residues occur at 26 positions along the sequences of α and β chains of hemoglobin and sperm whale myoglobin. Of these, 10 are hydrophobic residues that pack against the heme (Dickerson and Geis 1983).

Myoglobin stability depends on both tertiary structure and interactions of globin with heme. Once hemin is lost, protein unfolding occurs in two stages: (i) partial unfolding to an intermediate molten globule stage and (ii) complete unfolding of helices. In the intermediate molten globule form, the A, G, and H helices are intact, B and E helices are unfolded, and there is a complete loss of the heme binding site (Hughson, Wright and Baldwin 1990, Hargrove *et al.* 1994a). Hemin loss and apoglobin unfolding are not necessarily correlated, and in general, expression yields of recombinant myoglobins cannot be predicted by their rates of hemin loss. (Hargrove *et al.* 1994a).

Increasing the polarity of heme pocket residues at positions B10, E11, or G8 destabilizes tertiary structure, promotes globin unfolding, and inhibits expression (Hargrove *et al.* 1994a). However, these effects are independent of the stability of the heme-globin linkages: B10(Leu → Asn), E11(Val → Ser), and G8(Ile → Thr) have 10-15-fold higher rates of hemin loss, E11(Val → Asn) has a normal rate of hemin loss, and E11(Val → Thr) has a lower rate of hemin loss, and yet, all of these sperm whale mutant apomyoglobins are unstable. Conversely, increasing the hydrophobicity of the distal pocket stabilizes apomyoglobin. Introduction of a Phe residue at positions E11 or G8 increases apomyoglobin stability and has little effect on rates of hemin loss. Phe or Leu at distal position E7 also stabilizes apomyoglobin but at the expense of heme stability and function (Hargrove *et al.* 1994a) (see Table 8.2).

The stability of apomyoglobins is also species dependent (see Table 8.2). Pig and human apomyoglobins are less stable than sperm whale apomyoglobin even though their rates of hemin loss are similar at pH 7.0 (Hargrove *et al.* 1994a). The reasons for these differences are not yet fully understood.

8.6.4.1 Expression of Recombinant Hemoglobins

Stability of heme binding and protein folding are important not just for potential toxicities of hemoproteins or for long-term storage but also for the expression yields of recombinant hemoglobins. Globins expressed in nonfusion *E. coli* systems retain an extra methionine at the N-terminus due to insufficient post-translational processing (Hoffman *et al.* 1990, Hernan *et al.* 1992). The extra methionine alters the N-terminal conformation of hemoglobin (Kavanaugh, Rogers and Arnone 1992), and in myoglobin, it decreases the stability of wild-type protein moderately (Hargrove *et al.* 1994a) (see Table 8.2).

To process the N-terminal methionine of recombinant hemoglobin expressed in *E. coli*, Shen *et al.* (1993) constructed a plasmid in which human α and β globins are co-expressed with the *E. coli* methionine aminopeptidase gene. This expression system produces a recombinant hemoglobin that is identical to that of native human hemoglobin in high yield but with some altered (probably inverted) β-subunit heme conformation. Shen *et al.* were able to correct the altered conformational form by oxidizing the recombinant hemoglobin and then reducing it back to the Fe^{2+} state, perhaps by allowing the hemin to dissociate and then reorient in the proper conformation.

8.7 Heme Toxicity and Catabolism

Free heme is highly lipophilic and in the absence of heme-binding proteins, partitions into cellular membranes by intercalating between lecithin acyl chains (see Figure 8.2). The major barrier to heme translocation across the membrane is movement of the charged propionate groups through the apolar region of the bilayer (Light and Olson 1990).

8.7.1 Fenton Chemistry

Heme-induced membrane damage occurs by catalysis of peroxidation reactions with unsaturated lipids. Iron as either free iron or in heme catalyzes the Haber-Weiss reaction (Haber and Weiss 1934):

(5) $\qquad O_2^- \cdot + H_2O_2 \rightarrow O_2 + \cdot OH^- + \cdot OH$

The overall mechanism for this reaction includes the Fenton reaction (Eq. 6) in which interaction between Fe^{2+} and H_2O_2 creates a homolytic cleavage of the peroxide bond to produce Fe^{3+} and the hydroxyl radical ($\cdot OH$) (see Figure 8.2):

(6) $\qquad Fe^{2+} + H_2O_2 \rightarrow Fe^{3+} + \cdot OH^- + \cdot OH$

In the presence of superoxide, Fe^{3+} is reduced back to Fe^{2+} (Eq. 7), creating a highly reactive catalytic cycle:

(7) $\qquad Fe^{3+} + O_2^- \cdot \rightarrow Fe^{2+} + O_2$

It is not clear whether ferrous iron in the hemoglobin molecule can act as a Fenton reagent (Puppo and Halliwell 1988, Sadrzadeh *et al.* 1984), but even if it does, the highly reactive ·OH formed at the iron center would most likely react with nearby protein residues rather than escape to the solvent to oxidize lipids. Using sickle cell ghost membranes as a model system, Repka and Hebbel (1991) have shown that three iron compartments are capable of acting as catalysts of the Fenton reaction: (i) preexisting free iron, (ii) free iron released from hemoglobin by oxidative stress, and (iii) an iron compartment that is not chelatable by deferoxamine. This latter, unchelatable component is probably free heme.

In vitro, unmodified hemoglobin promotes the formation of species that are equivalent in reactivity to ·OH (Sadrzadeh *et al.* 1984). *In vivo*, the presence of hemoglobin increases oxidative injury in tissues of the central nervous system (Sadrzadeh *et al.* 1984), the kidney (Paller 1988), and the lungs (Seibert *et al.* 1991). Nevertheless, in both situations, it is probably iron or heme released from hemoglobin by oxidant stress that promotes the damage (Gutteridge 1986, Puppo and Halliwell 1988).

When heme is bound to haptoglobin or to hemopexin, it is prevented from participating in peroxidative catalysis, but when bound to albumin, it retains about half of its catalytic activity (Muller-Eberhard and Nikkilä 1989). The effect of cell-free, hemoglobin-based blood substitutes on these plasma proteins is still not well established. If they are saturated by acellular hemoglobin-based products, their protective effects will be lost. However, hemoglobin modified by cross-linking between α subunits does not quench haptoglobin fluorescence, suggesting that since it cannot dissociate, it cannot bind haptoglobin (Panter *et al.* 1994).

8.7.2 Effects of Heme on Endothelial Cells

In the presence of hemin, cultured endothelial cells become acutely susceptible to oxidant damage from activated inflammatory cells or soluble H_2O_2. Tin-substituted protoporphyrin or protoporphyrin minus the iron does not promote cellular damage. Thus, iron is necessary for the injury (Balla *et al.* 1991).

H_2O_2-mediated endothelial cytotoxicity occurs after acute exposure to ferric but not to ferrous hemoglobin, suggesting that release of hemin is the causative factor (Balla *et al.* 1993). There is a correlation between the rates of hemoglobin autoxidation and the levels of heme uptake and damage to cultured endothelial cells (Motterlini *et al.* 1995). With $\alpha\alpha$-cross-linked hemoglobin, both hemin loss and the production of free radical species are probably involved in cytotoxicity, because deferoxamine, an iron chelator and hydroxyl ion scavenger, limits cellular damage. In contrast, deferoxamine has no effect on the cytotoxicity induced by native hemoglobin, so that in this case, heme alone may be damaging (Motterlini *et al.* 1995).

It is interesting that chronic exposure of endothelium to hemin or to methemoglobin brings about a protective response against further oxidant stress. This occurs due to a rapid rise in cellular levels of heme oxygenase mRNA and a subsequent increase in cellular levels of ferritin, an iron sequestration protein (Balla *et al.* 1992, Balla *et al.* 1993).

8.7.3 Heme Catabolism by Heme Oxygenase

Heme oxygenase is a microsomal enzyme that catalyzes the rate-limiting step in heme catabolism in concert with NADPH-cytochrome P-450 reductase. Heme oxygenase is not itself a hemoprotein but becomes a transitory one after binding its substrate, heme, in a 1:1 molar ratio. Before metabolism, heme oxygenase-bound ferric heme is reduced by NADPH so that it binds O_2. Further reducing equivalents from NADPH convert the bound heme-O_2 complex to an α-hydroxyheme and, ultimately, to biliverdin. In the process, an α-methene carbon bridge is eliminated as CO, and iron is released (see Figure 8.2). In total, three molecules of oxygen and six reducing equivalents are required to degrade one molecule of heme. In mammals, biliverdin is converted to bilirubin by the enzyme biliverdin reductase (Abraham *et al.* 1988, Maines 1988).

8.7.3.1 *The Role of Heme Oxygenase in Cytoprotection*

Expression of heme oxygenase occurs in response to stress in cultured human and mammalian cell lines (Applegate, Luscher and Tyrrell 1991). It is now believed that heme oxygenase has an important secondary role to that of heme catabolism as a heat shock protein (hsp) to protect against oxidative damage (Maines 1988, Keyse and Tyrrell 1989). Two isozymes of heme oxygenase, HO-1 and HO-2, have been identified that are the products of different genes with separate mechanisms of regulation (Maines 1988). HO-1, which is hsp32, is induced by its substrate hemin (Stocker, Glazer and Ames 1987a), heat shock (Shibahara, Muller and Taguchi 1987), heavy metals (Maines and Kappas 1974), ultraviolet (UV) radiation and oxidative stress, (Keyse and Tyrrell 1989), and possibly during renal ischemia (see Maines *et al.* 1993 and Paller, Nath and Rosenberg 1993 for opposing results). In rats, mild restraint or surgical stress induces hsp's (Udelsman, Blake and Holbrook 1991, Udelsman *et al.* 1993), and 24-48 hours after surgery to implant catheters in Sprague-Dawley rats, levels of hepatic heme oxygenase activity are significantly increased (Motterlini *et al.*, unpublished data).

Three possible roles for the cytoprotective effects of heme oxygenase have been proposed: (i) Iron released during heme catabolism induces the production of ferritin (see Figure 8.2), decreasing the level of intracellular free iron that can be involved in iron-catalyzed free radical reactions (Vile and Tyrrell 1993, Eisenstein *et al.* 1991). (ii) Heme oxygenase rids the cell of high concentrations of free heme, which has pro-oxidant activity through its participation in the Fenton reaction. During sustained oxidant stress, catabolism of excess heme should restrict generation of hydroxyl radical and ferryl species. (iii) Bile reaction products of heme catabolism (*i.e.*, bilirubin) have strong antioxidant properties because they are scavengers of reactive oxygen metabolites (Stocker *et al.* 1987b).

The more direct roles of heme oxygenase in cytoprotection outlined in (ii) and (iii) above have not been proved. Even though cultured aortic endothelial cells pretreated with hemin (Balla *et al.* 1992) or human skin fibroblasts pre-irradiated with UV-A radiation (Vile *et al.* 1994) become resistant to further oxidative injury, the induction of ferritin in response to enhanced heme oxygenase activity is probably the major protective route through its depletion of labile pools of free iron. In a human breast cancer cell line exposed to cellular oxidants, heme oxygenase and bilirubin production alone could not constitute an antioxidant defense (Nutter, Sierra and Ngo 1994).

8.7.3.2 Regulation of Blood Pressure by Heme Oxygenase Activity

An important hemodynamic role of heme oxygenase has recently come to light. In spontaneously hypertensive rats, there is an inverse relation in the levels of heme oxygenase and cytochrome P-450 arachidonate hydroxylase, a heme protein that catalyzes the conversion of arachidonic acid to vasoactive metabolites (Da Silva *et al.* 1994). When the renal HO-1 gene is activated in this model by treatment with $SnCl_2$, an elevation in heme oxygenase activity occurs, cytochrome P-450 hydroxylase activity decreases, and blood pressure drops. The cytoplasmic level of active cytochrome P-450 is theorized to be regulated by the availability of intracellular heme, and the degradation of hepatic cytochrome P-450 by heme oxygenase has been demonstrated directly in rats (Kutty *et al.* 1988). Thus, increased heme oxygenase activity may have a negative effect on cytochrome P-450 activity, possibly leading to a decline in blood pressure.

Observations in rats exchanged transfused with αα-cross-linked hemoglobin are consistent with this theory. Preliminary experiments at UCSD show that the onset of the hypertensive response to αα-cross-linked hemoglobin depends on the time after surgery that the exchange transfusion is performed. At 1.5-hours or 7-days post-surgery, significant increases in mean arterial pressure are observed upon 50% exchange transfusion, but at 24-hours post-surgery, blood pressure remains constant (Motterlini *et al.*, unpublished data). These results correlate well with time courses of heme oxygenase synthesis and activity. In aortic endothelium exposed to hemin, the level of heme oxygenase mRNA rises between 2- and 4-hours post-exposure (Balla *et al.* 1992), and after surgery in rats, heme oxygenase levels are increased at 24 and 48 hours but return to normal after 4 days (Motterlini *et al.*, unpublished data).

The possibility exists that the high level of heme oxygenase activity at 24-hours post-surgery inhibits an increase in blood pressure. This effect is not observed at either 1.5-hours or 7-days post-surgery, because hypothetically, at the earlier time, significant heme oxygenase transcription has not occurred, and at the later time, heme oxygenase levels have increased and returned to normal. Whether this interpretation is valid is still open to study, but clearly, the role of heme oxygenase as a stress-response protein raises new questions about the interpretation of hemodynamic data.

8.8 Conclusions and Summary

The application of hemoglobin-based blood substitute products creates a situation in which large amounts of modified hemoproteins are introduced free into the circulation. In the absence of red cell reductive and antioxidant systems, there is little protection against oxidative by-products of hemoglobin denaturation. Oxidative reaction intermediates and heme loss present a catalytic environment for membrane damage through the peroxidation of lipids and proteins, and acute exposure of endothelial cells to methemoglobin or free hemin is cytotoxic. Thus, inhibiting autoxidation becomes a primary strategy in the design of blood substitutes.

Strategies to combat denaturation of hemoglobin-based products are evolving. Site-directed mutagenesis is a particularly effective tool for engineering hemoglobin because it allows fine-tuning of protein function and stability. At this point, myoglobin has provided a simple prototype for these experiments, and probably not surprising, only a few myoglobin mutants have been prepared to date that are promising in terms of stability (*e.g.*, low rates of autoxidation and hemin loss and improved or unchanged stability of the apoglobin). These are **B10(Leu → Phe)**, which has a 10-fold lower rate of autoxidation, a small decrease in the rate of hemin loss, and only a slight decrease in stability of the apoprotein, and **E11(Val → Phe)**, which has a slightly higher rate of autoxidation, no change in the rate of hemin dissociation, and a large increase in stability of the apoprotein. Compared with native sperm whale myoglobin, the O_2 affinity of the B10 mutant is 20-fold higher, and the O_2 affinity of the E7 mutant is slightly lower. It is still not clear whether lower O_2 affinities, lower rates of autoxidation, lower rates of hemin loss, or greater stability of the apoprotein should be favored in the design of a blood substitute, but it is highly likely that multiple mutations will be necessary to achieve the final desired properties, once they are defined.

It is generally believed that the O_2 affinity of a hemoglobin-based red cell substitute should be similar to that of red blood cells for adequate O_2 delivery to tissue. However, there is really little evidence to support this assumption. Considering the strong correlation between O_2 affinity and the rate of autoxidation, reappraisal of an optimal P50 is in order. There may be little advantage in enhancing O_2 off-loading if, in parallel, hemoglobin denaturation and oxidative damage are accelerated. There is the additional paradox in that hemin exposure to endothelium creates a long-term cytoprotective response, although, whether this protection offers an advantage after acute injury has occurred is debatable.

8.9 Acknowledgments

This work was supported by the National Heart, Lung and Blood Institute of the National Institutes of Health (P01 HL48018).

8.10 References

Abraham, N.G., H.-C. Lin, M.L. Schwartzman, R.D. Levere, and S. Shibahara. The physiological significance of heme oxygenase. *Int. J. Biochem.* 20: 543-558, 1988.

Alayash, A.I., J.C. Fratantoni, C. Bonaventura, J. Bonaventura, and E. Bucci. Consequences of chemical modification on the free radical reactions of human hemoglobin. *Arch. Biochem. Biophys.* 298: 114-120, 1992.

Applegate, L.A., P. Luscher, and R.M. Tyrrell. Induction of heme oxygenase: a general response to oxidant stress in cultured mammalian cells. *Cancer Res.* 51: 974-978, 1991.

Balla, G., H.S. Jacob, J. Balla, M. Rosenberg, K. Nath, F. Apple, J.W. Eaton, and G.M. Vercellotti. Ferritin: a cryoprotective antioxidant strategem of endothelium. *J. Biol. Chem.* 267: 18148-18153, 1992.

Balla, J., H.S. Jacob, G. Balla, K. Nath, J.W. Eaton, and G.M. Vercellotti. Endothelial cell heme uptake from heme proteins: induction of sensitization and desensitization to oxidant damage. *Proc. Natl. Acad. Sci. USA* 90: 9285-9289, 1993.

Balla, G., G.M. Vercellotti, U. Muller-Eberhard, J. Eaton, and H.S. Jacob. Exposure of endothelial cells to free heme potentiates damage mediated by granulocytes and toxic oxygen species. *Lab. Invest.* 64: 648-655, 1991.

Banerjee, R., and R. Cassoly. Preparation and properties of the isolated alpha and beta chains of human hemoglobin in the ferri state. Investigation of oxidation-reduction equilibria. *J. Mol. Biol.* 42: 337-349, 1969.

Banerjee, R., and J.-M. Lhoste. Nonequivalence of human hemoglobin chains in the oxidation-reduction and heme-transfer reactions. *Eur. J. Biochem.* 67: 349-356, 1976.

Benesch, R.E., and S. Kwong. The stability of the heme-globin linkage in some normal, mutant, and chemically modified hemoglobins. *J. Biol. Chem.* 265: 14881-14885, 1990.

Brantley, R.E., S.J. Smerdon, A.J. Wilkinson, E.W. Singleton, and J.S. Olson. The mechanism of autooxidation of myoglobin. *J. Biol. Chem.* 268: 6995-7010, 1993.

Brunori, M. Reactions of hemoglobin with O_2 and NO. *18th European Conference on Microcirculation*, Rome, Meeting Abstract, 1994.

Bunn, H.F., and B.G. Forget. *Hemoglobin: Molecular, Genetic and Clinical Aspects.* Philadelphia, W.B. Saunders Co., 1986.

Bunn, H.F., and J.H. Jandl. Exchange of heme among hemoglobins and between hemoglobin and albumin. *J. Biol. Chem.* 243: 465-475, 1968.

Cadenas, E. Biochemistry of oxygen toxicity. *Annu. Rev. Biochem.* 58: 79-110, 1989.

Carver, T.E., R.E. Brantley, E.W. Singleton, R.M. Arduini, M.L. Quillin, G.N. Phillips, and J.S. Olson. A novel site-directed mutant of myoglobin with an unusually high O_2 affinity and low autooxidation rate. *J. Biol. Chem.* 267: 14443-14450, 1992.

Chevalier, A., D. Guillochon, N. Nedjar, J.M. Piot, M.W. Vijayalakshmi, and D. Thomas. Glutaraldehye effect on hemoglobin: evidence for an ion environment modification based on electron paramagnetic resonance and Mössbauer spectroscopies. *Biochem. Cell Biol.* 68: 813-818, 1990.

Da Silva, J.-L., M. Tiefenthaler, E. Park, B. Escalante, M.L. Schwartzman, and N.G. Abraham. Tin-mediated heme oxygenase gene activation and cytochrome P-450 hydroxylase inhibition in spontaneously hypertensive rats. *Am. J. Med. Sci.* 307: 173-181, 1994.

De Baere, I., M.F. Perutz, L. Kiger, M.C. Marden, and C. Poyart. Formation of two hydrogen bonds from the globin to the heme-linked oxygen molecule in *Ascaris* hemoglobin. *Proc. Natl. Acad. Sci. USA* 91: 1594-1597, 1994.

Demma, L.S., and J.M. Salhany. Subunit inequivalence in superoxide anion formation during photooxidation of human oxyhemoglobin. *J. Biol. Chem.* 254: 4532-4535, 1979.

Dickerson, R.E., and I. Geis. *Hemoglobin: Structure, Function, Evolution, and Pathology.* Menlo Park: Benjamin/Cummings Publishing Co., 1983.

Dix, T.A., R. Fontana, A. Panthani, and L.J. Marnett. Hematin-catalyzed epoxidation of 7,8-dihydroxy-7,8-dihydrobenzo[a]pyrene by polyunsaturated fatty acid hydroperoxides. *J. Biol. Chem.* 260: 5358-5365, 1985.

Eisenstein, R.S., D. Garcia-Mayol, W. Pettingell, and H.N. Munro. Regulation of ferritin and heme oxygenase synthesis in rat fibroblasts by different forms of iron. *Proc. Natl. Acad. Sci. USA* 88: 688-692, 1991.

Fermi, G., and M.F. Perutz. *Haemoglobin and Myoglobin.* Oxford: Clarendon Press, 1981.

Giulivi, C., and K.J.A. Davies. Hydrogen peroxide-mediated ferrylhemoglobin generation *in vitro* and in red blood cells. *Meth. Enzymol.* 231: 490-496, 1994.

Guillochon, D., L. Esclade, and D. Thomas. Effect of glutaraldehyde on hemoglobin: oxidation-reduction potentials and stability. *Biochem. Pharmacol.* 35: 317-323, 1986.

Gutteridge, J.M.C. Iron promoters of the Fenton reaction and lipid peroxidation can be released from haemoglobin by peroxides. *FEBS Lett.* 201: 291-295, 1986.

Haber, F., and J. Weiss. The catalytic decomposition of hydrogen peroxide by iron salts. *Proc. Roy. Soc. Lond. Ser. A* 147: 332-351, 1934.

Hargrove, M.S., S. Krzywda, A.J. Wilkinson, Y. Dou, M. Ikeda-Saito, and J.S. Olson. Stability of myoglobin: a model for the folding of heme proteins. *Biochemistry* 33: 11767-11775, 1994a.

Hargrove, M.S., E.W. Singleton, M.L. Quillin, L.A. Ortiz, G.N. Phillips, J.S. Olson, and A.J. Mathews. His64(E7) → Tyr apomyoglobin as a reagent for measuring rates of hemin dissociation. *J. Biol. Chem.* 269: 4207-4214, 1994b.

Hernan, R.A., H.L. Hui, M.E. Andracki, R.W. Noble, S.G. Sligar, J.A. Walder, and R.Y. Walder. Human hemoglobin expression in *Escherichia coli*: importance of optimal codon usage. *Biochemistry* 31: 8619-8628, 1992.

Hoffman, S.J., D.L. Looker, J.M. Roehrich, P.E. Cozart, S.L. Durfee, J.L. Tedesco, and G.L. Stetler. Expression of fully functional tetrameric human hemoglobin in *Escherichia coli*. *Proc. Natl. Acad. Sci. USA* 87: 8521-8525, 1990.

Honig, G.R., L.N. Vida, B.B. Rosenblum, M.F. Perutz, and G. Fermi. Hemoglobin Warsaw (Phe-β42 (CD1) → Val), an unstable variant with decreased oxygen affinity. *J. Biol. Chem.* 265: 126-132, 1990.

Hughson, F.M., P.E. Wright, and R.L. Baldwin. Structural characterization of a partly folded apomyoglobin intermediate. *Science* 249: 1544-1548, 1990.

Jacob, H., and K. Winterhalter. Unstable hemoglobins: the role of heme loss in Heinz body formation. *Proc. Natl. Acad. Sci. USA* 65: 697-701, 1970.

Kaca, W., R. Roth, K.D. Vandegriff, G.C. Chen, F.A. Kuypers, and J. Levin. Effects of bacterial endotoxin on human crosslinked and native hemoglobins, submitted, 1995.

Kannan, R., R. Labotka, and P.S. Low. Isolation and characterization of the hemichrome-stabilized membrane protein aggregates from sickle erythrocytes. *J. Biol. Chem.* 263: 13766-13773, 1988.

Kavanaugh, J.S., P.H. Rogers, and A. Arnone. High-resolution x-ray study of deoxy recombinant human hemoglobins synthesized from β-globins having mutated amino termini. *Biochem.* 31: 8640-8647, 1992.

Keyse, S.M., and R.M. Tyrrell. Heme oxygenase is the major 32-kDa stress protein induced in human skin fibroblasts by UVA radiation, hydrogen peroxide, and sodium arsenite. *Proc. Natl. Acad. Sci. USA* 86: 99-103, 1989.

Kindt, J.T., A. Woods, B.M. Martin, R.J. Cotter, and Y. Osawa. Covalent alteration of the prosthetic heme of human hemoglobin by Br-CCl_3. *J. Biol. Chem.* 267: 8739-8743, 1992.

Kutty, R.K., R.F. Daniel, D.E. Ryan, W. Levin, and M.D. Maines. Rat liver cytochrome P-450b, P-420b, and P-420c are degraded to biliverdin by heme oxygenase. *Arch. Biochem. Biophys.* 260: 638-644, 1988.

Levy, A., K. Alston, and J.M. Rifkind. Dynamics of hemoglobin investigated by Mössbauer spectroscopy. *J. Biomol. Struct. Dyn.* 1: 1299-1309, 1984.

Light, W.R., and J.S. Olson. Transmembrane movement of heme. *J. Biol. Chem.* 265: 15623-15631, 1990.

Macdonald, V.W., K.D. Vandegriff, R.W. Winslow, D. Currell, C. Fronticelli, J. C. Hsia, and J.C. Bakker. Oxidation rates and stability in solution of mono- and bivalently cross-linked human hemoglobin. *Biomater. Artif. Cells Immobil. Biotech.* 19: 425, 1991.

Maines, M.D. Heme oxygenase: function, multiplicity, regulatory mechanisms, and clinical applications. *FASEB J.* 2: 2557-2568, 1988.

Maines, M.D., and A. Kappas. Cobalt induction of hepatic heme oxygenase with evidence that cytochrome P-450 is not essential for this enzyme activity. *Proc. Natl. Acad. Sci. USA* 71: 4293-4297, 1974.

Maines, M.D., R.D. Mayer, J.F. Ewing, and W.K. McCoubrey. Induction of kidney heme oxygenase-1 (HSP32) mRNA and protein by ischemia/reperfusion: possible role of heme as both promotor of tissue damage and regulator of HSP32. *J. Pharmacol. Exp. Therapeu.* 264: 457-462, 1993.

Mansouri, A., and K.H. Winterhalter. Nonequivalence of chains in hemoglobin oxidation and oxygen binding. Effect of organic phosphates. *Biochemistry* 13: 3311-3314, 1974.

Motterlini, R., R. Foresti, K.D. Vandegriff, M. Intaglietta, and R.M. Winslow. Heme- and iron-induced oxidative stress in vascular endothelial cells exposed to acellular hemoglobins. *Am. J. Physiol.*, submitted, 1995.

Muller-Eberhard, U., J. Javid, H.H. Liem, A. Hanstein, and M. Hanna. Plama concentrations of hemopexin, haptoglobin, and heme in patients with various hemolytic diseases. *Blood* 32: 811-815, 1968.

Muller-Eberhard, U., and H. Nikkilä. Transport of tetrapyrroles by proteins. *Semin. Hematol.* 26: 86-104, 1989.

Nutter, L.M., E.E. Sierra, and E.O. Ngo. Heme oxygenase does not protect human cells against oxidant stress. *J. Lab. Clin. Med.* 123: 506-514, 1994.

Osawa, Y., J.F. Darbyshire, C.A. Meyer, and A.I. Alayash. Differential susceptibilities of the prosthetic heme of hemoglobin-based red cell substitutes. *Biochem. Pharmacol.* 46: 2299-2305, 1993.

Osawa, Y., and K. Korzekwa. Oxidative modification by low levels of HOOH can transform myoglobin to an oxidase. *Proc. Natl. Acad. Sci. USA* 88: 7981-7085, 1991.

Paller, M.S. Hemoglobin- and myoglobin-induced acute renal failure in rats: role of iron in nephrotoxicity. *Am. J. Physiol.* 253: F539-544, 1988.

Paller, M.S., K.A. Nath, and M.E. Rosenberg. Heme oxygenase is not expressed as a stress protein after renal ischemia. *J. Lab. Clin. Med.* 122: 341-345, 1993.

Panter, S.S., K.D. Vandegriff, P.O. Yan, and R.F. Regan. Assessment of hemoglobin-dependent neurotoxicity: alpha-alpha crosslinked hemoglobin. *Artif. Cells, Blood Substitutes, Immobil. Biotech.* 22: 399-413, 1994.

Puppo, A., and B. Halliwell. Formation of hydroxyl radicals from hydrogen peroxide in the presence of iron. Is haemoglobin a biological Fenton reagent? *Biochem. J.* 249: 185-190, 1988.

Quillin, M.L., R.M. Arduini, J.S. Olson, and G.N. Phillips. High-resolution crystal structures of distal histidine mutants of sperm whale myoglobin. *J. Mol. Biol.* 234: 140-155, 1993.

Repka, T., and R.P. Hebbel. Hydroxyl radical formation by sickle erythrocyte membranes: role of pathologic iron deposits and cytoplasmic reducing agents. *Blood* 78: 2753-2758, 1991.

Rifkind, J.M., O. Abugo, A. Levy, and J. Heim. Detection, formation, and relevance of hemichromes and hemochromes. *Meth. Enzymol.* 231: 449-480, 1994.

Romero-Herrera, A.E., M. Goodman, H. Dene, D. Bartnicki, and H. Mizukami. An exceptional amino acid replacement on the distal side of the iron atom in proboscidean myoglobin. *J. Mol. Evol.* 17: 140-147, 1981.

Sadrzadeh, S.M.H., E. Graf, S.S. Panter, P.E. Hallaway, and J.W. Eaton. Hemoglobin: a biological Fenton reagent. *J. Biol. Chem.* 259: 14354-14356, 1984.

Seibert, A.F., A.E. Taylor, J.B. Bass, and J. Haynes. Hemoglobin potentiates oxidant injury in isolated rat lungs. *Am. J. Physiol.* 260: H1980-H1984, 1991.

Shaanan, B. Structure of human oxyhaemoglobin at 2.1Å resolution. *J. Mol. Biol.* 171: 31-59, 1983.

Shen, T.-J., N.T. Ho, V. Simplaceanu, M. Zou, B.N. Green, M.F. Tam, and C. Ho. Production of unmodified human adult hemoglobin in *Escherichia coli. Proc. Natl. Acad. Sci. USA* 90: 8108-8112, 1993.

Shibahara, S., S. Muller, and H. Taguchi. Transcriptional control of rat heme oxygenase by heat shock. *J. Biol. Chem.* 262: 12889-12892, 1987.

Shikama, K. A controversy on the mechanism of autooxidation of oxymyoglobin and oxyhaemoglobin: oxidation, dissociation, or displacement. *Biochem. J.* 223: 279-280, 1984.

Smerdon, S.J., G.G. Dodson, A.J. Wilkinson, Q.H. Gibson, R.S. Blackmore, T.E. Carver, and J.S. Olson. Distal pocket polarity in ligand binding to myoglobin: structural and functional characterization of a threonine68 E(11) mutant. *Biochemistry* 30: 6252-6260, 1991.

Smith, M.L., K. Hjortsberg, P.-H. Romeo, J. Rosa, and K.-G. Paul. Mutant hemoglobin stability depends upon location and nature of single point mutation. *FEBS Lett.* 169: 147-150, 1984.

Springer, B.A., K.D. Egeberg, S.G. Sligar, R.J. Rohlfs, A.J. Mathews, and J.S. Olson. Discrimination between oxygen and carbon monoxide and inhibition of autooxidation by myoglobin. *J. Biol. Chem.* 264: 3057-3060, 1989.

Stocker, R., A.N. Glazer, and B.N. Ames. Antioxidant activity of albumin-bound bilirubin. *Proc. Natl. Acad. Sci. USA* 84: S918-S922, 1987a.

Stocker, R., Y. Yamamoto, A.F. McDonagh, A.N. Glazer, and B.N. Ames. Bilirubin is an anitoxidant of possible physiological importance. *Science* 235: 1043-1046, 1987b.

Tucker, P.W., S.E.V. Phillips, M.F. Perutz, R. Houtchens, and W.S. Caughey. Structure of hemoglobins Zürich [His(63)β replaced by Arg] and Sydney [Val(67)β replaced by Ala] and role of the distal residues in ligand binding. *Proc. Natl. Acad. Sci. USA* 75: 1076-1080, 1978.

Udelsman, R., M.J. Blake, and N.J. Holbrook. Molecular response to surgical stress: specific and simultaneous heat shock protein induction in the adrenal cortex, aorta, and vena cava. *Surgery* 110: 1125-1131, 1991.

Udelsman, R., M.J. Blake, C.A. Stagg, D. Li, D.J. Putney, and N.J. Holbrook. Vascular heat shock protein expression in response to stress. *J. Clin. Invest.* 91: 465-473, 1993.

Vandegriff, K.D., and Y.C. Le Tellier. A comparison of rates of heme exchange: site-specifically cross-linked *versus* polymerized human hemoglobins. *Artif. Cells, Blood Subsitutes, Immobil. Biotech.* 22: 443-455, 1994.

Vile, G.F., S. Basu-Modak, C. Waltner, and R.M. Tyrrell. Heme oxygenase 1 mediates an adaptive response to oxidative stress in human skin fibroblasts. *Proc. Natl. Acad. Sci. USA* 91: 2607-2610, 1994.

Vile, G.F., and R.M. Tyrrell. Oxidative stress resulting from ultraviolet A irradiation of human skin fibroblasts leads to heme oxygenase-dependent increase in ferritin. *J. Biol. Chem.* 268: 14678-14681, 1993.

Vincent, S.H. Oxidative effects of heme and porphyrins on proteins and lipids. *Semin. Hematol.* 26: 105-113, 1989.

Wallace, W.J., R.A. Houtchens, J.C. Maxwell, and W.S. Caughey. Mechanism of autooxidation for hemoglobins and myoglobin. Promotion of superoxide production by protons and anions. *J. Biol. Chem.* 257: 4966-4977, 1982.

Weiss, J.J. Nature of the iron-oxygen bond in oxyhaemoglobin. *Nature* 202: 83-84, 1964.

Wilks, A., and P.R. Ortiz de Montellano. Intramolecular translocation of the protein radical formed in the reaction of recombinant sperm whale myoglobin with H_2O_2. *J. Biol. Chem.* 267: 8827-8833, 1992.

Winterbourn, C.C. Oxidative denaturation in congenital hemolytic anemias: the unstable hemoglobins. *Semin. Hematol.* 27: 41-50, 1990.

Yamamoto, Y., and G.N. La Mar. ^{1}H NMR study of dynamics and thermodynamics of heme rotational disorder in native and reconstituted hemoglobin A. *Biochemistry* 25: 5288-5297, 1986.

Zhang, L., A. Levy, and J.M. Rifkind. Autoxidation of hemoglobin enhanced by dissociation into dimers. *J. Biol. Chem.* 266: 24698-24701, 1991.

Chapter 9

Red Cell Substitutes in the Kidney

Roland C. Blantz, M.D., Andrew P. Evan, Ph.D.* and
Francis B. Gabbai, M.D.

*Department of Medicine, University of California, San Diego and
Veterans Affairs Medical Center, 3350 La Jolla Village Drive (9111-H),
San Diego, California 92161*

**Department of Anatomy, University of Indiana, Indiana University
Medical Center, Indianapolis, Indiana 46202-5120*

9.1 Introduction

The utility of cross-linked hemoglobin (Hgb) solutions has been predicted to be limited in part by potential renal toxicity and consequent renal dysfunction (Hess *et al.* 1989, Keipert and Chang 1987, Messmer 1978, Snyder *et al.* 1987). However, this concern has been based upon models and assumptions that may not be entirely pertinent since direct studies have not as yet been performed in both euvolemic and hypovolemic models pertinent to the utility of red cell substitutes. Some of the potential concerns that have been voiced in the literature are as follows:

1. Filtered Hgb products may result in tubular damage analogous to the clinical syndrome of rhabdomyolysis in which myoglobin is filtered at the glomerulus along with other cellular products (Fischereder, Trick and Nath 1994). Obviously, the clinical settings are somewhat different. It has been demonstrated that rhabdomyolysis is a form of acute renal failure which is associated with distal tubular obstruction and some proximal tubular damage, both events not necessarily by the myoglobin per se. There are some recent data, however, that do suggest that the endothelial derived nitric oxide system is significantly perturbed in rhabdomyolysis in that arginine levels decrease and that the syndrome is ameliorated, in part, by providing arginine, suggesting that myoglobin may consume nitric oxide leading to renal vasoconstriction (Brezis *et al.* 1991).

Blood Substitutes: Physiological Basis of Efficacy
Winslow et al., Editors
© Birkhäuser Boston 1995

2. Hemoglobin by interaction with endothelium or other circulating monocyte macrophage systems may activate a variety of cytokines and elaborate materials which cause tissue damage, *i.e.*, TNF or other cytokines (Lieberthal *et al.* 1987, Xia *et al.* 1993).

3. The fact that Hgb binds nitric oxide may lead to renal vasoconstriction to the degree that nitric oxide serves as a counter-balancing vasodilatory influence within the kidney, or alternatively, that free Hgb may somehow enhance the activity of other major vasoconstrictor systems, *i.e.*, adrenergic and/or angiotensin II (Nishi *et al.* 1994, Gabbai *et al.* 1993).

4. Extracellular Hgb and/or the liberation of heme and free iron may somehow enhance the formation of reactive oxygen species (Paller and Nath 1991).

5. Acute renal failure may be the consequence of a more generalized process affecting the kidney only secondarily as part of this overall process.

Information will be presented which, in part, answers some of these concerns. Prior to these specific retorts, some background information is required on the specific, unique characteristics expressed in the kidney.

9.2 Oxygen Environment Within The Kidney

It is assumed that many forms of acute renal failure derived from the consequences of ischemia or at least relative oxygen and substrate deficiency related to the demands of the organ. The kidney is the most efficient autoregulatory organ in the body and regulates blood flow across wide ranges of systemic blood pressure, utilizing both myogenic and tubuloglomerular feedback systems, the latter of which relates reabsorption to filtration rate (Thurau and Schnermann 1965, Thomson and Blantz 1993). Blood flow to the kidneys constitutes up to 20% of cardiac output, providing the kidney the highest flow per gram of tissue of any organ. However, this large supply of renal blood flow is not necessary for large substrate and oxygen requirements but rather related to the fact that the kidney is an organ that produces approximately 150 liters of protein-free ultrafiltrate each day (Smith 1937). The major oxygen requirements of the kidney are linked to reabsorption, and oxygen consumption relates directly to NaCl transport. If filtration is prevented, oxygen demands decrease precipitously, therefore, ischemia and substrate deficiency in the kidney should be defined in terms of its normal workload (filtration rate and reabsorption is normal or near normal).

The arteriovenous oxygen difference in the kidney is very small, suggesting that oxygen requirements of the kidney are relatively modest compared to oxygen supply. It has been estimated that only 5% of normal oxygen delivery is required to prevent ischemia, but these are under conditions in which glomerular filtration no longer occurs, therefore, reabsorptive demands are minimal.

The fact that there is a low A-Vo$_2$ difference in the kidney might suggest indirectly that the intrarenal oxygen environment is quite rich. In fact, recent studies have suggested that this is not the case. The density of vessels in the kidney is quite high with each glomerulus exhibiting continuous flow and arteriolar and venular structure closely packed (Smith 1937). Recent investigations using PO$_2$ electrodes implanted within the kidney cortex and medulla have defined the local PO$_2$ and pCO$_2$ (Brezis *et al.* 1991, Schurek *et al.* 1990, Schurek and Johns 1990). The pCO$_2$ in the cortex is on the order of 60-65 mm Hg, primarily as a consequence of the high rate of proton secretion required for bicarbonate reabsorption. The pCO$_2$ in the medulla is much lower since a lesser rate of acidification proton secretion occurs within this region. Alternatively, the PO$_2$ in the cortex remains much lower than arterial PO$_2$ at approximately 60-65 mm Hg. Studies by Schurek *et al.* utilizing PO$_2$ electrodes demonstrate that when arterial PO$_2$ is raised to 550 mm Hg, the cortical PO$_2$ remains in the range of 60-65 mm Hg (Schurek *et al.* 1990). The PO$_2$ in the outer medulla is even lower and appears to be sensitive to variations in flow, exhibiting values of PO$_2$ from 25 mm to 38 mm Hg (Brezis *et al.* 1991). These latter results suggest that the outer medulla provides a zone of borderline hypoxia, suggesting that much of the ATP generated for transport functions might be anaerobic of glycolytic in origin rather than through normal, highly efficient aerobic mechanisms. These PO$_2$ values oscillate within their ranges, corresponding to the frequency of alterations in pressure and flow within tubular structures, suggesting that PO$_2$ oscillations reflect variations in flow-dependent, solute or NaCl-NaHCO$_3$ transport (Schurek and Johns 1990).

How does the PO$_2$ remain so low with the cortex and medulla of the kidney in spite of high values for systemic arterial PO$_2$? This is due in part to a countercurrent exchange mechanism which derives from the fact that arteriolar and venular capillaries are densely packed within the kidney, promoting a preglomerular oxygen diffusion shunt among incoming and outgoing vessels, maintaining the cortex and medulla of the kidney at a relatively constant but relatively hypoxic condition (Schurek *et al.* 1990). Therefore, the A-Vo$_2$ difference in the kidney is quite low, and one might expect that variations in the oxygen and substrate delivery and/or transport demand could place the kidney under conditions of borderline ischemia, such that the form and magnitude of oxygen delivery could be critical to renal function.

9.3 Experimental Studies

9.3.1 Interactions of Cross-linked Hemoglobins and the Kidney - Studies in the Isolated Perfused Organ

Comments will be confined to interactions between cross-linked hemoglobin solutions and the kidney. Most of these studies have been performed in the cell-free, isolated perfused kidney which exhibits a rather high rate of renal plasma flow at normal perfusion pressures of approximately 30 ml/min, approximately ten fold above the normal for plasma flow rate (Lieberthal *et al.* 1987, Gabbai *et al.* 1994). This is primarily the consequence of very low renal vascular resistances, in part, related to high shear rate dependent rates of nitric oxide generation as well as the absence of the rheologic and NO inactivating properties of red blood cells. We demonstrated in the isolated perfused kidney that the low vascular resistances were not entirely the result of nitric oxide production since large doses of L-NMMA, a competitive inhibitor of nitric oxide synthase, does not normalize renal vascular resistance and flow (Gabbai *et al.* 1993). L-NMMA only decreases flow by approximately 20%. When one adds red blood cells at a normal hematocrit of 40%, renal plasma flow approaches normal at ~4-6 ml/min, and with red cells added the filtration fraction of the kidney approaches normal at approximately 20%, suggesting that red cells alone supply some element, possibly beyond the rheologic properties of red cells, that may increase renal vascular resistance beyond consumption of nitric oxide (Gabbai *et al.* 1994).

However, we have performed studies that suggest that the addition of Hgb (in 0.5g% increments) does yield increases in renal vascular resistance, utilizing cross-linked DBBF Hgb (Blantz *et al.* submitted 1994, Blantz, Peterson and Winslow 1992). In the cell-free, isolated perfused kidney preparation at 3g%, renal plasma flow rate decreases by approximately 20% which is comparable to values observed with the administration of L-NMMA, suggesting that nitric oxide consumption does increase renal vascular resistance in this preparation, and that the difference supplied by red cells is not totally the consequence of nitric oxide consumption. Alternatively, cyanomethemoglobin, a hemoglobin that does not acutely bind and inactivate nitric oxide, exerts no effects at 3g% on renal vascular resistance and flow in the isolated, cell-free perfused organ (Blantz *et al.* submitted 1994).

In addition to nitric oxide consumption and inactivation by cross-linked Hgb, Lieberthal and colleagues have suggested that thromboxane generation might contribute to the vasoconstriction in the isolated perfused kidney (Lieberthal, LaRaia and Valeri 1992). Both stroma-free and cross-linked Hgb decrease renal plasma flow and GFR. Approximately 50% of this vasoconstriction can be ameliorated by thromboxane synthase and receptor blockers. The exact origin of increased thromboxane generation was not known but it might suggest that cross-linked Hgb interaction with the endothelium may be critical to this event. More recent studies by Nishi and coworkers in the isolated perfused kidney preparation have examined the effects of L-NMMA and free Hgb on the reactivity of the isolated perfused kidney to both vasodilators and vasoconstrictors (Nishi *et al.* 1994). L-NMMA surprisingly produced very little modification of the effects of the vasoconstrictors angiotensin II and norepinephrine. However, cross-linked Hgb greatly magnified the effects of both angiotensin II and adrenergic agonists, suggesting that the effects of Hgb on the isolated perfused kidney cannot be attributed entirely to diminished nitric oxide activity and that Hgb, by mechanisms unknown, enhance the effect of a variety of naturally occurring vasoconstrictors. Given these observations in the isolated organ it would suggest that vascular resistance alterations upon exposure to cross-linked Hgb in the plasma are only partially a consequence of nitric oxide inactivation, and that free Hgb may exert independent effects upon vascular resistance. However, it should be recalled that these *in vitro* studies are short term in nature, usually terminating within 60-90 minutes, and may not be pertinent to the systemic effects of exposure to large volume cross-linked Hgb.

9.3.2 Studies Utilizing Large Volume Hgb Isovolemic Exchange Studies in the Awake Animal

We have utilized the chronically catheterized rat preparation for a variety of pathophysiologic studies in the past and have recently applied this model to determine how well 50% isovolemic cross-linked Hgb exchange is tolerated in the awake animal and whether renal dysfunction is observed (Blantz *et al.* submitted 1994, Blantz, Peterson and Winslow 1992). The advantages of this preparation are that it does not require anesthesia and the stress of surgical preparation is distant from the event. The animals are prepared for surgery at least a week prior to the initiation of the experiment and are trained to sit quietly in a non stressed condition. After control observations of mean arterial pressure, GFR and renal plasma flow, animals were submitted to 50% isovolemic exchange with cross-linked Hgb solutions at 7g% concentration. In addition, we also examined the effects of isovolemic exchange with albumin and with the addition of L-NMMA in the two protein exchange groups. The important observations of these studies were as follows:

1.   GFR and renal plasma flow remained constant for several hours after exchange and through 5 days, a time at which hematocrit had been restored to normal values by the normal production of red blood cells (Blantz *et al.* submitted 1994, Blantz, Peterson and Winslow 1992).

2.   The addition of L-NMMA to either albumin or Hgb exchange groups did not greatly alter any of the renal or systemic parameters

(Blantz *et al.* submitted 1994). The reduction in hematocrit may have neutralized the effect of L-NMMA.

3.   It is clear that inhibition of nitric oxide synthase provides no added impact on Hgb exchange on blood pressure which was modestly elevated after isovolemic exchange. The importance of these results lies in the absolute constancy of GFR and renal plasma flow rates, in spite of the fact that there was modest hemoglobinuria. Approximately 3% of the Hgb administered was in the noncross-linked form and led to significant Hgb filtration. Most importantly, these studies in awake animals contrast strikingly when compared to studies in the acute surgically prepared and anesthetized animal.

9.3.3 Large Volume Cross-linked Hgb Isovolemic Exchange in the Acute Surgically Prepared Anesthetized Rat

We have studied a variety of acute surgically prepared animals who have then been submitted to either 50% albumin or 50% cross-linked Hgb exchange (Blantz *et al.* submitted 1994, Blantz, Peterson and Winslow 1992). Measurements have been frequent but periodic utilizing clearance and micropuncture techniques for approximately 200 minutes after exchange. There were several observations of a morphologic and molecular biologic nature. Animals submitted to either albumin or Hgb exchange exhibited no morphologic alterations of their endothelial cells derived from glomeruli using scanning electron microscopy (Figure 9.1).

Fenestrated glomerular epithelial cells were examined in several animals, and no alterations in endothelial cells and in the size of fenestrae were observed 200 minutes following 50% isovolemic Hgb exchange. In addition, we have examined the expression of cytokine inducible nitric oxide synthase in a variety of organs 120-200 minutes after exchange with albumin, cross-linked Hgb and in saline administered control. Endotoxin contamination was significant in standard commercially derived albumin preparations. However, endotoxin was barely detectable in the Hgb and saline control solutions. We observed significant inducible NOS mRNA in lung, spleen, kidney and other organs in the albumin exchange group. It is notable that there was no renal dysfunction or morbidity noted in this group, as will be observed later. Cross-linked Hgb was endotoxin-free but also caused increased mRNA for iNOS in both lung and spleen. The exact mechanism for this increased transcription of inducible NOS is not known, but endotoxin is not a candidate since solutions were endotoxin free. It remains possible, however, that Hgb is either directly activating cytokine release through interaction with monocyte-macrophage systems, or somehow nitric oxide consumption by Hgb leads indirectly to new transcription for the inducible enzyme.

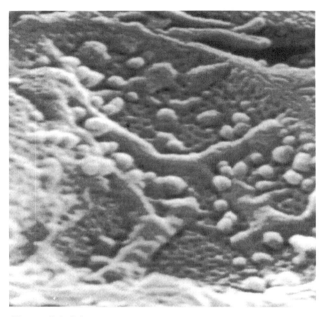

Figure 9.1. (a)

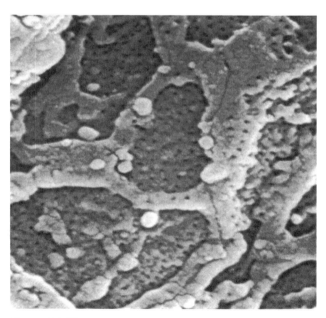

Figure 9.1 (b)

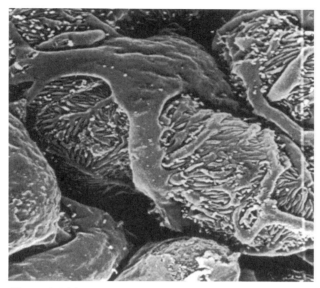

Figure 9.1 (c)

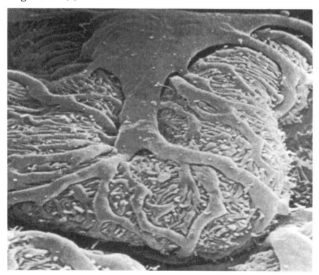

Figure 9.1 (d)

Figure 9.1 We have utilized scanning electron microscopy to examine both glomerular endothelial and epithelial cells after 50% albumin or αα cross-linked Hgb exchange (endothelial cells, panels a and b, and epithelial cells, panels c and d). The fenestrae were completely normal in size, and endothelial cells were normal in configuration in both albumin and Hgb groups. The glomerular visceral epithelial cells were also normal and equal to control animals after both albumin (panels a and c) and Hgb exchange (panels b and d) (endothelial cells - magnification X 42,000 and glomerular epithelial cells X 13,000).

Mortality was quite high in the acutely surgically prepared animals submitted to 50% isovolemic Hgb exchange. This stands in stark contrast to the results in the awake animal. Animals submitted to isovolemic albumin exchange survived completely with no evidence of renal dysfunction. In the surgically prepared animal, addition of L-NMMA, a nitric oxide synthase inhibitor, further magnified the mortality rate to nearly 100% and shortened the time of survival to 80-120 minutes after Hgb exchange. Arginine infusion may have provided some modest benefit by reducing mortality from 60 to 40%. Animals died from generalized cardiovascular collapse. Animals submitted to albumin exchange, even if treated with L-NMMA, also survived and exhibited nearly normal renal function.

Studies on cardiac outputs were performed in order to provide some insights into the mechanisms for cardiovascular instability (Blantz *et al.* submitted 1994). In animals injected with 3 differently labeled microspheres, cardiac output increased significantly 30 minutes after isovolemic Hgb exchange but returned to normal values at approximately 100-120 minutes after exchange. In addition, renal blood flow increased transiently as it did in the micropuncture studies but gradually fell to normal values at 100-120 minutes. The exact mechanism for the later cardiovascular collapse in this situation remains to be fully delineated. However, it does appear that acute surgical preparation or the stress of surgery somehow conditions the animal to high mortality conditions after isovolemic Hgb exchange. It is quite possible that the stress of surgery may induce enzyme systems that somehow interact with free Hgb in a coordinate fashion, leading to cardiovascular instability and death. Although binding of nitric oxide to Hgb may exert some acute effects on blood pressure and other cardiovascular parameters, the net effect of isovolemic exchange is renal vasodilation and increases in renal plasma flow and GFR, even in the anesthetized, surgically prepared preparation. The toxicities observed in the anesthetized preparation and especially animals submitted to acute surgery suggest that prior stress may be a preconditioning event which leads to increased morbidity and mortality after cross-linked Hgb administration. Further studies are required to delineate whether this is further magnified by clinical conditions in which red blood cell substitutes are to be utilized.

9.4 Acknowledgments

This work was supported by the NIH (P01 HL48018) and the Veterans Affairs Research Service.

9.5 References

Blantz, R.C., O.W. Peterson, and R.M. Winslow. Effects of 50% blood exchange with purified $\alpha\alpha$ cross-linked hemoglobin (αHb). *Clin. Res.* 40(2): 290A, 1992.

Blantz, R.C., O.W. Peterson, B.J. Tucker, L. Feng, C.B. Wilson, R.M. Winslow, and F.B. Gabbai. Functional assessments of 50% whole blood exchange with $\alpha\alpha$ cross-linked hemoglobin in awake and anesthetized, surgically manipulated rats. Submitted to *Blood*, 1994.

Brezis, M., S.N. Heyman, D. Dinour, F.H. Epstein, and S. Rosen. Role of nitric oxide in renal medullary oxygenation. Studies in isolated and intact rat kidneys. *J. Clin. Invest.* 88: 390, 1991.

Fischereder, M., W. Trick, and K.A. Nath. Therapeutic strategies in the prevention of acute renal failure. *Semin. Nephrol.* 14: 41-52, 1994.

Gabbai, F.B., O.W. Peterson, C.B. Wilson, and R.C. Blantz. Glomerular hemodynamics in the cell free (IPK) and erythrocytes perfused (IEPK) isolated perfused kidney. *J. Am. Soc. Nephrol.* 4(3): 578, 1993.

Gabbai, F.B., O.W. Peterson, S. Khang, C.B. Wilson, and R.C. Blantz. Glomerular hemodynamics in the cell-free and erythrocyte-perfused isolated rat kidney. *Am. J. Physiol.* 267: F423-F427, 1994.

Hess, J.R., S.O. Fadare, L.S.L. Tolentino, N.R. Bangal, and R.M. Winslow. The intravascular persistence of crosslinked human hemoglobin. *The Red Cell: Seventh Ann Arbor Conf.*, 351-360, 1989.

Keipert, P.E. and T.M.S. Chang. Effects of partial and total isovolemic exchange transfusion in fully conscious rats using pyridoxylated polyhemoglobin solution as a colloidal oxygen-delivering blood replacement fluid. *Vox. Sang.* 53: 7-14, 1987.

Lieberthal, W., E.F. Wolf, E.W. Merrill, N.G. Levinsky, and C.R. Valeri. Hemodynamic effects of different preparations of stroma-free hemolysates in the isolated perfused rat kidney. *Life Sci.* 41: 2525-2533, 1987.

Lieberthal, W., J. LaRaia, and C.R. Valeri. Role of thromboxane in mediating the intrarenal vasoconstriction induced by unmodified stroma-free hemoglobin in the isolated perfused rat kidney. *Biomat. Artif. Cells Immobil Biotech.* 20: 663-667, 1992.

Messmer, K. Hemodilution. *Surg. Clin. North Am.* 55: 659-678, 1978.

Nishi, K., S. Ueda, D. Sato, and K. Nishi. Effects of pyridoxalated hemoglobin polyoxyethylene conjugate and stroma-free hemoglobin on renal vascular responsiveness to vasoactive substances in isolated perfused rat kidney. *Artif. Organs* 18: 429-438, 1994.

Paller, M.S., and K.A. Nath. Radical thoughts about kidneys. *J. Lab. Clin. Med.* 118: 415-417, 1991.

Schurek, H.-J., and O. Johns. Tubuloglomerular feedback prevents nephron oxygen deficiency via TAL-segments. *J. Am. Soc. Nephrol.* 1: 603A, 1990.

Schurek, H.-J., U. Jost, H. Baumgartl, H. Bertram, and U. Heckman. Evidence for a preglomerular oxygen diffusion shunt in rat renal cortex. *Am. J. Physiol.* 259: F910, 1990.

Smith, H.W. *The Physiology of the Kidney*, Oxford, New York, 1937.

Snyder, S.R., E.W. Welby, R.Y. Walder, T.A. Williams, and J.A. Walder. HbXL99α: A hemoglobin derivative that is cross-linked between the α subunits is useful as a blood substitute. *Proc. Natl. Acad. Sci. USA* 84: 7280-7284, 1987.

Thomson, S.C., and R.C. Blantz. Homeostatic efficiency of tubuloglomerular feedback in hydropenia, euvolemia, and acute volume expansion. *Am. J. Physiol.* 264: F930, 1993.

Thurau, K., and J. Schnermann. Die Natrium-konzentration an den Macula Densa Zellen als regulierender Faktor fur das Glomerulumfiltrat (Mikropunktionsversuche), *Klin. Wochenschr.* 43: 410, 1965.

Xia, Y., L. Geng, T. Yoshimura, and C.B. Wilson. LPS-induced MCP-1, IL-1ß, and TNFα, mRNA expression in isolated erythrocyte-perfused rat kidney. *Am. J. Physiol.* 264: F774-F780, 1993.

Chapter 10

A Theoretical Analysis of Oxygen Transport: A New Strategy for the Design of Hemoglobin-Based Red Cell Substitutes

Kim D. Vandegriff, Ph.D. and Robert M. Winslow, M.D.

Department of Medicine, School of Medicine, University of California, San Diego, Veterans Affairs Medical Center (111-E), 3350 La Jolla Village Drive, San Diego, California 92161

ABSTRACT

In vitro experiments and experiments with artificial capillary systems suggest that transfer of oxygen from cell-free carriers to tissue sites may be faster than it is from red blood cells. On first consideration, this appears to be advantageous, but new data obtained by direct measurements in the microcirculation indicate that facilitated release may actually lead to an autoregulatory decrease in capillary perfusion. This leads to the conclusion that the well-known hemoglobin-oxygen equilibrium curve may not be adequate by itself to describe oxygen delivery by cell-free hemoglobin-based red cell substitutes to achieve optimal tissue oxygenation.

10.1 Introduction

For cell-free hemoglobins to be used in O_2-carrying resuscitation fluids, changes in the molecule are required that are made either through site-directed mutagenesis or chemical modification techniques. The primary goals of hemoglobin modification so far have been to prevent tetramer dissociation and to alter the O_2 binding affinity to mimic that of red blood cells *in vivo*. These protein engineering strategies have been successful primarily through cross-linking hemoglobin tetramers, which reduces dimer filtration in the kidneys and prevents nephrotoxicity. In the absence of renal dysfunction, the single most problematic physiologic effect is hemoglobin-induced vasoactivity, which has been observed with

Blood Substitutes: Physiological Basis of Efficacy
Winslow et al., Editors
© Birkhäuser Boston 1995

most hemoglobin solutions and manifests as hypertension and increased peripheral vascular resistance.

The advantage of red cell substitutes as resuscitation colloids over conventional plasma expanders is their potential for providing O_2 transport in addition to oncotic activity. However, because of their vasoactivity, even with high O_2 capacity and O_2 affinity similar to that of whole blood, hemoglobin-based O_2 carriers have not always augmented O_2 transport to tissue above that achieved with either albumin or dextran solutions. For example, in a shock pig model, augmented blood O_2 content resulting from resuscitation with hemoglobin was offset by increased vascular resistance and low cardiac output (Hess, Macdonald and Brinkley 1993). Similar findings have been reported in the dog (Rooney, Hirsch and Mathru 1993) and in cat neurovasculature (Ulatowski *et al.* 1993). In the microcirculation of a hamster skin-window preparation, the increased O_2 content carried by $\alpha\alpha$-cross-linked hemoglobin was offset by vasoconstriction in arterioles and decreased functional capillary density (Tsai, Kerger and Intaglietta 1995).

O_2 is delivered to tissue primarily through transport in the microcirculation. Thus, loss of functional capillaries becomes particularly detrimental to tissue oxygenation. Capillaries are thought to be elastic rather than contractile and to have the mechanical properties of their surrounding tissue so that the size of their lumen is determined by the hydrodynamic pressure imposed by larger arteriolar vessels upstream (Fung, Zweifach and Intaglietta 1966, Tsai, Kerger and Intaglietta 1995). The diameter of the larger arteriolar vessels upstream, on the other hand, is controlled directly by the tone of the vessel smooth muscle, and when these larger vessels constrict, capillary flow is impeded. Under normal conditions, the number of perfused capillaries is uniform, but when this system is perturbed, like it appears to be with the $\alpha\alpha$-cross-linked hemoglobin solution, tissue oxygenation suffers. The decrease in tissue PO_2 that occurs because of the loss of capillaries is exacerbated by a marked increase in O_2 consumption across the arteriolar vessel wall, and as a result, tissue PO_2 is decreased in half (Tsai, Kerger and Intaglietta 1995). It can be inferred from these studies that the potential benefit of hemoglobin-based O_2 carriers is limited, at this time, by their vasoactivity.

One mechanism by which hemoglobin-induced vasoconstriction may occur is through inhibition of endothelial-derived nitric oxide (EDNO), either by direct binding of NO as a hemoglobin ligand or by superoxide inactivation of NO to form the peroxynitrite anion (Ignarro 1990). This remains controversial, and there are additional conflicting results regarding the effects of O_2 tension on the action of NO. In one study, an inverse relation was found between the saturation of hemoglobin and inhibition of vasodilation caused by NO (Iwamoto and Morin 1993). In another study, blocking EDNO in canine skeletal muscle had no affect on hypoxia-induced vasodilation, suggesting an uncoupling between tissue O_2 levels and NO action (Vallet *et al.* 1994).

A second mechanism that has not been fully explored in investigations of blood substitutes is that hemoglobin-induced vasoactivity may simply be the result of an autoregulatory hemodynamic response to the unique O_2 transporting properties of these solutions. Guyton originally suggested that when diffusion of O_2 to tissue exceeded O_2 demands, an autoregulatory vasoconstrictor response would return tissue O_2 levels to normal (Guyton *et al.* 1964). This theory for a hemodynamic response to hyperoxia has been tested in dogs in which the O_2 affinity of intraerythrocytic hemoglobin was decreased by adding inositol hexaphosphate, and consistent with Guyton's theory, cardiac output fell and total peripheral resistance increased (Liard and Kunert 1993). Also consistent with the theory are observations in skeletal muscle microcirculation where a correlation has been found between the ambient PO_2 exposed to tissue, the number of perfused capillaries, and arteriolar diameter. As PO_2 increased, arterioles constricted and functional capillary density decreased (Lindbom, Tuma and Arfors 1980).

10.2 O_2 Equilibrium Binding of Red Blood Cells *versus*

Hemoglobin Solutions

The traditional view that shifting the O_2 equilibrium curve to the right (higher P50) facilitates O_2 unloading in tissue sites has been shown in a number of experimental models using red blood cells (Malmberg, Hlastala and Woodson 1979, Martin *et al.* 1979, Woodson *et al.* 1982) and in clinical settings (Collins 1980). However, in certain situations, the opposite may be true. For example, Barcroft *et al.* (1923) believed that increased O_2 affinity (lower P50) was important in adaptation to high altitude. Their reasoning was based on analogy with the placental circulation in which fetal blood has a higher affinity for O_2 than that of the mother. Furthermore, modification of hemoglobin to increase its affinity has been shown to convey superior survival in hypoxic rats (Eaton *et al.* 1974), and mutant hemoglobins with increased O_2 affinity may precondition subjects to hypoxia (Hebbel *et al.* 1978). On the summit of Mt. Everest, the P50 of whole blood is about 19 torr (Winslow, Samaja and West 1984), similar to that of fetal hemoglobin, which suggests that lowering the O_2 affinity of red blood cells would not improve O_2 transport in situations where pulmonary uptake of O_2 is diffusion limited (*i.e.*, low arterial O_2 partial pressure at high altitude).

It has been generally assumed that the O_2 equilibrium curve for hemoglobin-based blood substitutes must be similar to or right-shifted from that of red blood cells for adequate O_2 delivery to tissues. Outside of the cell, human hemoglobin loses its primary effector, 2,3-

diphosphoglycerate, and O_2 affinity is increased. While the native hemo-globin structure has been modified in the design of blood substitutes to decrease O_2 affinity for suitable O_2 off-loading, the primary assumption about O_2 affinity does not take into account diffusional factors that limit O_2 transport by red blood cells *in vivo*. The position and shape of the O_2 equilibrium curve are important determinants of O_2 delivery, but the physical transport of O_2 also depends upon diffusion from the red cells to tissue.

According to a theory advanced by Homer, Weathersby and Kiesow (1981) and Federspiel and Popel (1986), the low plasma solubility of O_2 creates a major barrier to diffusion of O_2 in the plasma spaces between red cells and between red cells and capillary walls. In contrast, acellular hemoglobin-based O_2 carriers are dispersed homogeneously throughout the plasma, which eliminates these diffusion barriers and, theoretically, accelerates O_2 transport to tissue (Figure 10.1). Based on these theoreti-cal concepts and new experimental data from the microcirculation model (Tsai, Kerger and Intaglietta 1995), it becomes necessary to consider that the vasoconstriction induced by cell-free hemoglobin might be an autoregulatory response to accelerated O_2 delivery to vasoactive blood vessels. A hypothetical mechanism for this effect, which has theoretical and experimental support, is presented below.

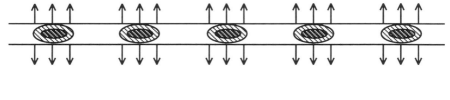

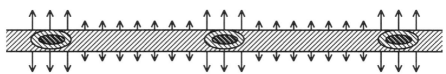

Figure 10.1 A theoretical analysis of the particulate nature of blood on oxygen re-lease in capillaries (Federspiel and Popel 1986). Red blood cells alone are shown on top. Red blood cells plus hemoglobin solution are shown at the bottom.

10.3 Kinetic Measurements of O_2 Exchange by Cellular

versus Acellular Hemoglobins

In 1927, Hartridge and Roughton first discovered that O_2 uptake by red blood cells was approximately 40 times slower than O_2 uptake by a com-parable cell-free hemoglobin solution (Hartridge and Roughton 1927) (Figure 10.2). Their results were confirmed later by Roughton (1959) and other investigators, including Coin and Olson (1979), Weingarden, Mizukami and Rice (1982a,b), Kagawa and Mochizuki (1982), and Vandegriff and Olson (1984a,b,c), who used more modern experimental techniques. This later work demonstrated conclusively that the lower rates

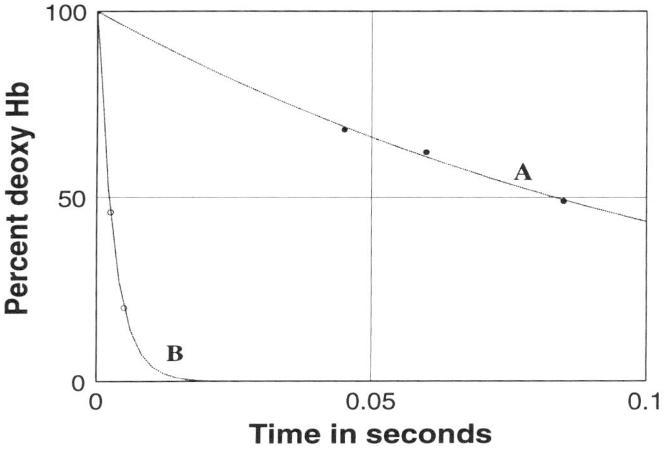

Figure 10.2 Time courses for oxygen uptake by (**A**) intact and (**B**) lysed sheep red blood corpuscles. Adapted from Hartridge and Roughton (1927).

of O_2 uptake and release by intact cells are due to diffusional barriers surrounding red blood cells that are absent with acellular hemoglobin solutions. About half of the lowering in the rate in a rapid-mixing apparatus is due to intracellular O_2 diffusion, and the other half to extracellular resistance to O_2 diffusion through unmixed solvent layers surrounding flowing red blood cells (Coin and Olson 1979, Vandegriff and Olson 1984a), results that are entirely consistent with Homer, Weathersby, and Kiesow's (1981) hypothesis.

Quantitatively, the resistance to O_2 diffusion through extracellular plasma layers *in vivo* is less well defined, but in the microcirculation, plasma layers between red cells and capillary walls in muscle tissue are on the order of 0.5-1 μm. These are distances with significant effects on experimental rates of O_2 uptake (Vandegriff and Olson 1984a). Even greater plasma diffusion distances have been observed in liver blood vessels (Cliff 1976).

10.4 Fick's Law of Diffusion

The Fick diffusion equation has been used in a simple boundary analysis to estimate the external resistance to O_2 diffusion for red blood cells (Nicolson and Roughton 1951, Kagawa and Mochizuki 1982, Vandegriff and

Olson 1984a,c). The O_2 flux at the red cell interface can be calculated using a diffusion gradient for PO_2 values across a diffusion barrier (P_1 - P_2),

$$(1) \quad \frac{\delta[O_2]}{\delta t} \simeq \frac{D(P_1 - P_2)}{\Delta r}$$

where D is the oxygen diffusion constant between phases P_1 and P_2, and Δr is the diffusion distance between P_1 and P_2.

Using this analysis, we can consider the effects on O_2 transport. In the capillaries of the lung (Figure 10.3), the rate of O_2 uptake by red blood

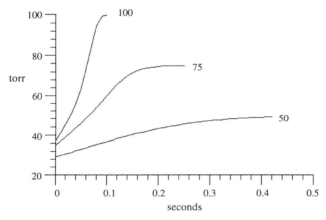

Figure 10.3 Diffusion limitation of O_2 uptake by red blood cells. Calculations are for a 70 kg person with 5 ml/kg/min O_2 uptake and normal pulmonary function. The three curves correspond to three levels of alveolar O_2 tension, 100, 75, and 50 torr. The length of each line corresponds to the time required for arterial PO_2 to reach equilibrium with alveolar gas.

cells is determined by the high partial pressure of O_2 in alveolar gas spaces (*i.e.*, P_1 ~100 torr in Eq. 1) because the O_2 diffusion gradient to red blood cells is steep. Under normal conditions in the lung (PAO_2 ~100 torr), the exact O_2-binding characteristics of cellular hemoglobin, which determines P_2 in this case, are less important. This, however, becomes invalid when the O_2 affinity of hemoglobin is very low, because then the concentration of O_2 free in plasma (*i.e.*, P_2) rises, making the O_2 gradient more narrow. Alternatively, if alveolar O_2 is low (*e.g.*, breathing air at high altitude), the diffusion gradient also would be more narrow, but in this case due to a decrease in P_1.

For O_2 release to contracting muscle tissue, the diffusion gradient is in the opposite direction. The O_2 concentration near the capillary surface remains close to zero because of highly active mitochondria (*i.e.*, P_2 ~0 torr). The rate of O_2 efflux at the red cell surface, or for extracellular hemoglobin solutions at the capillary surface, becomes more dependent on the chemical reaction between hemoglobin and O_2. Since P_2 ~0, the gradient driving diffusion (P_1 - P_2) becomes directly proportional to the partial pressure of O_2 in equilibrium with hemoglobin, which depends on the O_2 affinity of hemoglobin (or more simply, the P50).

10.5 Experimental Results of O_2 Exchange *In Vitro*

These theoretical analyses have been used successfully to explain *in vitro* measurements of O_2 uptake and release by cell-free hemoglobin solutions, liposome-encapsulated hemoglobin, and red blood cells in rapid-mixing experiments (Farmer *et al.* 1989, Vandegriff and Olson 1984c) and by cell-free hemoglobin solutions and intact red blood cells in an artificial capillary system (Boland *et al.* 1987, Lemon *et al.* 1987).

Kinetics of O_2 release in a rapid-mixing experiment are shown in Figure 10.4. The O_2 affinities of the cell-free and encapsulated hemoglobins

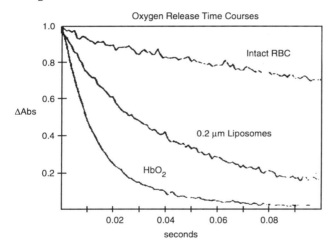

Figure 10.4 Time courses for O_2 release by human red blood cells, liposome-encapsulated hemoglobin, and stroma-free hemoglobin at 25°C and pH 7.4. Internal heme concentrations were 20 mM for the red blood cells and 6 mM for the liposomes. Adapted from Farmer *et al.* 1989.

used in this experiment were ~3-fold higher than the O_2 affinity of the red blood cells (*i.e.*, there were no hemoglobin effectors co-encapsulated in the liposomes) (Vandegriff and Olson 1984c). At the same total heme concentration, O_2 release from the cell-free hemoglobin was faster than from either the liposome-encapsulated hemoglobin or red blood cells, even though the O_2 affinities of the cell-free and liposome-encapsulated hemoglobins were the same and significantly higher than for the red blood cells. Thus, when hemoglobin is encapsulated, O_2 release is slower because O_2 must diffuse out of the vesicle to the site of consumption. Furthermore, even though the liposomes had higher O_2 affinity than that of the red blood cells, O_2 release from the liposomes was still faster for two reasons: (1) the liposomes had a lower internal heme concentration (6 *versus* 20 mM), and (2) they were smaller (*i.e.*, 0.2 μm in diameter compared to 8 μm for red blood cells), providing a much smaller surface area to volume ratio for diffusion.

For O_2 exchange in the artificial capillary, the following has been observed: (i) At equivalent heme concentrations, the rates of O_2 uptake and O_2 release are higher for hemoglobin solutions than for red blood cells. (ii) O_2 affinity has little influence on rates of O_2 uptake, but (iii) rates of O_2 release are proportional to the P50 of the oxygen carriers (Lemon *et al.* 1987, Boland *et al.* 1987).

The overall conclusions from this work are that the rate of O_2 uptake by a hemoglobin solution is largely independent of O_2 affinity but, because of the absence of significant diffusion barriers, higher than that for red blood cells. For both acellular and cellular oxygen carriers, the rate of O_2 release is proportional to O_2 affinity, but even at the same P50, O_2 release from hemoglobin solutions is faster because, while diffusion is still limiting, there is less of a barrier to diffusion. For example, even with a P50 1.5 times lower than that of a red cell suspension, O_2 release from an extracellular hemoglobin solution was faster (Figure 10.5) (Boland *et al.* 1987).

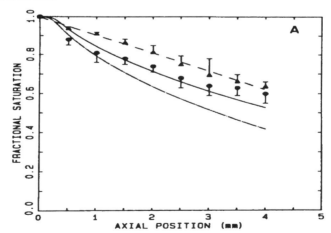

Figure 10.5 Time courses of O_2 release by red blood cell suspensions (triangles, P50 = 20 mm Hg) and hemoglobin solutions (circles, P50 = 13 mm Hg) of the same hemoglobin content at 37°C and pH 7.4. The solid and dashed lines are theoretical simulations for the hemoglobin solution and red blood cell suspension, respectively. The dotted-dashed line is a theoretical curve for a hemoglobin solution with a P50 of 20 mm Hg. Adapted from Boland *et al.* 1987.

10.6 Conclusions

The traditional view that O_2 transport to tissue can be enhanced by lowering the O_2 affinity of hemoglobin may be overly simplistic to describe O_2 transport by cell-free O_2 carriers. To create an effective blood substitute, O_2 must be delivered to tissue in the face of decreased red cell mass and physiologic compensatory mechanisms. The goal should be to design a solution that provides an O_2 flux at the blood vessel wall similar to that provided by red blood cells, because altered transport at the vessel wall may stimulate autoregulatory responses that change the balance between O_2 delivery and consumption. In the case of acellular hemoglobin solutions with O_2 binding properties similar to those of red blood cells, these regulatory responses may actually overcompensate for accelerated O_2 delivery at the precapillary level, causing vasoconstriction and increased O_2 consumption by the vessel wall.

According to the simple barrier analysis of Eq. 1, accelerated O_2 delivery to tissues by hemoglobin solutions with O_2 affinities similar to that of red blood cells occurs because of the decreased diffusion distance, Δr. To offset the smaller value for Δr, the O_2 diffusion gradient $(P_1 - P_2)$ must also be decreased to maintain an optimal value for the ratio $(P_1 - P_2)/\Delta r$. Since hemoglobin's affinity for O_2 influences primarily rates of O_2 release, it should be possible to alter these properties to manipulate O_2 release to tissue where O_2 consumption is high $(P_2 \sim 0)$. Theoretically, this can be accomplished by increasing the O_2 affinity of hemoglobin or by changing the shape of the binding curve so that less O_2 is unloaded at a given tissue PO_2.

The predicted consequences of such a strategy are as follows: At higher O_2 affinity (lower P50), less O_2 will dissociate from hemoglobin at any given PO_2, making less O_2 available to the plasma at the precapillary level. This may help to prevent hyperoxia upstream from capillary beds to avoid autoregulatory vasoconstriction. As a result, capillary flow would be maintained and tissue oxygenation would be preserved.

10.7 Acknowledgments

This work was supported by the National Heart, Lung and Blood Institute of the National Institutes of Health (P01 HL48018).

10.8 References

Barcroft J., C.A. Binger, A.V. Bock, J.H. Doggart, H.S. Forbes, G. Harrop, J.C. Meakins, and A.C. Redfield. Observations upon the effect of high altitude on the physiological processes of the human body carried out in the Peruvian Andes chiefly at Cerro de Pasco. *Philos. Trans. Roy. Soc. Lond. Ser. B* 211: 351-480, 1923.

Boland, E.J., P.K. Nair, D.D. Lemon, J.S. Olson, and J.D. Hellums. An *in vitro* capillary system for studies on microcirculatory oxygen transport. *J. Appl. Physiol.* 62: 791-797, 1987.

Cliff, W.J. *Blood Vessels.* Cambridge: Cambridge University Press, 1976.

Coin, J.T., and J.S. Olson. The rate of oxygen uptake by human red blood cells. *J. Biol. Chem.* 254: 1178-1190, 1979.

Collins, J.A. Abnormal hemoglobin-oxygen affinity and surgical hemotherapy. *Bibl. Haematol.* 46: 59-69, 1980.

Eaton, J.W., T.D. Skelton, and E. Berger. Survival at extreme altitude: protective effect of increased hemoglobin-oxygen affinity. *Science* 185: 743-744, 1974.

Farmer, M.C., A.S. Rudolph, K.D. Vandegriff, M.D. Hayre, S.A. Bayne, and S.A. Johnson. Liposome-encapsulated hemoglobin: oxygen binding properties and respiratory function. In *Blood Substitutes* (T.M.S. Chang and R.P. Geyer, eds.) New York: Marcel Dekker, 1989, pp. 289-299.

Federspiel, W.J., and A.S. Popel. A theoretical analysis of the effect of the particulate nature of blood on oxygen release in capillaries. *Microvasc. Res.* 32: 164-189, 1986.

Fung, Y.C., B.W. Zweifach, and M. Intaglietta. Elastic environment of the capillary bed. *Circ. Res.* 19: 441-461, 1966.

Guyton, A.C., J.M. Ross, O. Carrier, and J.R. Walker. Evidence for tissue oxygen demand as the major factor causing autoregulation. *Circ. Res.* 14: 1-60, 1964.

Hartridge, H., and F.J.W. Roughton. The rate of distribution of dissolved gases between the red blood corpuscle and its fluid environment. Part I. Preliminary experiments on the rate of uptake of oxygen and carbon monoxide by sheep's corpuscles. *J. Physiol.* 62: 232-242, 1927.

Hebbel, R.P., J.W. Eaton, R.S. Kronenberg, and E.D. Zanjane. Human llamas. Adaptation to altitude in subjects with high hemoglobin-oxygen affinity. *J. Clin. Invest.* 62: 593-600, 1978.

Hess, J.R., V.W. Macdonald, and W.W. Brinkley. Systemic and pulmonary hypertension after resuscitation with cell-free hemoglobin. *J. Appl. Physiol.* 74: 1769-1778, 1993.

Homer, L.D., P.K. Weathersby, and L.A. Kiesow. Oxygen gradients between red blood cells in the microcirculation. *Microvasc. Res.* 22: 308-323, 1981.

Ignarro, L.J. Biosynthesis and metabolism of endothelium-derived nitric oxide. In *Annu. Rev. Pharmacol. Toxicol.* (R. George, A.K. Cho and T.F. Blaschki, eds.) Palo Alto: Annual Reviews, Inc., 1990, pp. 535-560.

Iwamoto, J., and F.C. Morin. Nitric oxide inhibition varies with hemoglobin saturation. *J. Appl. Physiol.* 75: 2332-2336, 1993.

Kagawa, T., and M. Mochizuki. Numerical solution of partial differential equations describing oxygenation rates of the red blood cell. *Jap. J. Physiol.* 32: 197-218, 1982.

Lemon, D.D., P.K. Nair, E.J. Boland, J.S. Olson, and J.D. Hellums. Physiological factors affecting oxygen transport by hemoglobin in an *in vitro* capillary system. *J. Appl. Physiol.* 62: 798-806, 1987.

Liard, J.F., and M.P. Kunert. Hemodynamic changes induced by low blood oxygen affinity in dogs. *Am. J. Physiol.* 264: R396-R401, 1993.

Lindbom, L., R.F. Tuma, and K.-E. Arfors. Influence of oxygen on perfused capillary density and capillary red cell velocity in rabbit skeletal muscle. *Microvasc. Res.* 19: 197-208, 1980.

Malmberg, P.O., M.P. Hlastala, and R.D. Woodson. Effect of increased blood-oxygen affinity on oxygen transport in hemorrhagic shock. *J. Appl. Physiol.* 47: 889-895, 1979.

Martin, J.L., M. Duvelleroy, B. Teisseire, and M. Duruble. Effect of an increase in HbO_2 affinity on the calculated capillary recruitment of an isolated rat heart. *Pflügers Arch.* 382: 57-61, 1979.

Nicolson, P., and F.J.W. Roughton. A theoretical study of the influence of diffusion and chemical reaction velocity on the rate of exchange of carbon monoxide and oxygen between the red blood corpuscle and the surrounding fluid. *Proc. Roy. Soc. Lond. Ser. B* 138: 241-264, 1951.

Rooney, M.W., L.J. Hirsch, and M. Mathru. Hemodilution with oxyhemoglobin. Mechanism of oxygen delivery and its superaugmentation with a nitric oxide donor (sodium nitroprusside). *Lab. Invest.* 79: 60-72, 1993.

Roughton, F.J.W. Diffusion and simultaneous chemical reaction velocity in haemoglobin solutions and red cell suspensions. In *Progress in Biophysics and Biophysical Chemistry* (J.A.V. Butler and B. Katz, eds.) London: Pergamon Press, 1959, pp. 55-104.

Tsai, A.G., H. Kerger, and M. Intaglietta. Microcirculatory consequences of blood substitution with αα-hemoglobin. In *Blood Substitutes: Physiological Basis of Efficacy* (R.M. Winslow, K.D. Vandegriff and M. Intaglietta, eds.) Boston: Birkhäuser, 1995, pp. 155-174.

Ulatowski, J.A., M.A. Williams, T. Nishikawa, R.C. Koehler, R.J. Traystman, and E. Bucci. Maintenance of cerebral oxygen transport with bovine crosslinked hemoglobin. *Blood Substitutes and Related Products*, Philadelphia, September 21-22, Meeting Abstract, 1993.

Vallet, B., S.E. Curtis, M.J. Winn, C.E. King, C.K. Chapler, and S.M. Cain. Hypoxic vasodilation does not require nitric oxide (EDRF/NO) synthesis. *J. Appl. Physiol.* 76: 1256-1261, 1994.

Vandegriff, K.D., and J.S. Olson. A quantitative description in three dimensions of oxygen uptake by human red blood cells. *Biophys. J.* 45: 825-835, 1984a.

Vandegriff, K.D., and J.S. Olson. The kinetics of O_2 release by human red blood cells in the presence of external sodium dithionite. *J. Biol. Chem.* 259: 12609-12618, 1984b.

Vandegriff, K.D., and J.S. Olson. Morphological and physiological factors affecting oxygen uptake and release by red blood cells. *J. Biol. Chem.* 259: 12619-12627, 1984c.

Weingarden, M., H. Mizukami, and S.A. Rice. Transient effects on the initial rate of oxygenation of red blood cells. *Bull. Math. Biol.* 44: 119-134, 1982a.

Weingarden, M., H. Mizukami, and S.A. Rice. Factors defining the rate of oxygen uptake by the red blood cell. *Bull. Math. Biol.* 44: 135-147, 1982b.

Winslow, R.M., M. Samaja, and J.B. West. Red cell function at extreme altitude on Mount Everest. *J. Appl. Physiol.* 56: 109-116, 1984.

Woodson, R.D., J.H. Fitzpatrick, D.J. Costello, and D.D. Gilboe. Increased blood oxygen affinity decreases canine brain oxygen consumption. *J. Lab. Clin. Med.* 100: 411-424, 1982.

Chapter 11

Microcirculatory Consequences of Blood Substitution with αα-Hemoglobin

Amy G. Tsai, Ph.D., Heinz Kerger, M.D., and Marcos Intaglietta, Ph.D.

Department of Bioengineering, University of California, San Diego, La Jolla, California 92093-0412

ABSTRACT

Blood substitution with non-cellular fluids lowers blood viscosity and increases blood flow velocity, altering the oxygen delivering capacity of the microcirculation and the balance between diffusive and convective oxygen transport. When blood volume is maintained with either crystalloid or colloidal volume replacement, microcirculatory oxygen transport is adequate for red blood cell losses of the order of 60%. Hemoglobin solutions utilizing αα-hemoglobin show microvascular effects similar to those found with non-oxygen carrying solutions, but are reduced and the additional intrinsic oxygen carrying capacity of the blood/solution mixture is not manifested. Mathematical analysis of microvascular data shows that these effects are due to increased oxygen delivery capacity of the circulation, which increases arteriolar oxygenation. Ensuing metabolic autoregulatory responses lead to an oxygenation paradox resulting in vasoconstriction, impairment of functional capillary density, and increased microcirculatory metabolism, causing decreased tissue oxygenation. It is proposed that higher viscosity, left-shifted hemoglobin oxygen saturation solutions will fully oxygenate tissue.

11.1 Introduction

Blood substitution or replacement leads to the dilution of the red cell mass, and if water oxygen solubility is the only oxygen carrying mechanism of the replacement fluid, intrinsic oxygen carrying capacity of the resulting mixture decreases in proportion to the lowered hematocrit.

Blood Substitutes: Physiological Basis of Efficacy
Winslow et al., Editors
© Birkhäuser Boston 1995

This is not necessarily negative, since very low hematocrits, corresponding to losses of the red blood cell mass of the order of 70%, can be survived with relative ease. Conversely, our ability to compensate for comparatively smaller losses of blood volume is limited. A 30% deficit in blood volume can lead to irreversible shock if not rapidly corrected.

Maintenance of normovolemia is the objective of all forms of blood substitution or replacement. The ensuing dilution of blood constituents, termed hemodilution, produces systemic and microvascular phenomena underlying all forms of blood replacement. Fluids available to accomplish volume restitution can be broadly classified as crystalloid solutions, colloid solutions and oxygen carrying solutions. Each of these, when introduced in the circulation, yields a mixture of original blood and substitute whose transport properties must provide adequate tissue oxygenation, a phenomenon taking place in the microvasculature. It follows that altered blood compositions should be analyzed in terms of systemic effects and how these, coupled with the altered transport properties of the circulating fluid, influence transport microcirculatory function.

11.2 Macro and Micro Blood Flow in Hemodilution

11.2.1 Viscosity Effects

When hemodilution maintains isovolemic conditions, changes in transport properties of the circulation are circumscribed to the alteration of blood viscosity and intrinsic oxygen carrying capacity. This condition results from using either crystalloid or colloidal solutions, providing a physiological reference with which to compare blood mixtures obtained with substitutes.

Blood viscosity in the major vessels is the parameter most strongly influenced by hemodilution, since in this compartment it is approximately proportional to the hematocrit squared and inversely proportional to shear rate (Quemada 1978). The pressure drop in the major vessel (arteries and veins) accounts for a relatively small portion of the viscous losses in the circulation, therefore the critical viscosity effect is that of the remainder of the circuit, namely the microcirculation.

In the microcirculation blood viscosity is linearly related to hematocrit (Lipowsky, Usami and Chien 1980) whereby hemodilution causes a directly proportional decrease in vascular resistance. If the heart were a constant pressure output pump, lowered viscosity would translate in a proportional increase in cardiac output and therefore flow velocity in the microcirculation. An additional lowering of viscosity takes place since increased flow velocity increases shear rate, on which blood viscosity is inversely dependent. While this is borne out by clinical and experimental studies, there are two additional competing mechanisms that contribute in setting cardiac output, namely: 1) Lowered viscosity in the venous circulation improves cardiac filling, contractility and therefore cardiac output, and 2) increased cardiac output coupled with lowered blood viscosity increases oxygen delivery capacity, triggering metabolic autoregulation and the increase of peripheral vascular resistance (Intaglietta 1989).

Changes in plasma viscosity due to the use of colloidal solutions appears to be a minor factor in the range of concentrations intended for blood substitutes. One of the highest colloidal concentration contemplated for use is 15% hemoglobin solutions (15 g/100 ml). The viscosity of such a solution can be evaluated from the equation due to Einstein (1956) which is exact for spherical particles up to a concentration of 1%:

$$(1) \qquad \eta_s = \eta_o(1 + 2.5c)$$

where η_s is the viscosity of the suspension, η_o is the viscosity of the suspending medium (water) and c is the volumetric concentration of particles. For a concentration of 15% this equation underestimates the viscosity by a factor of $6.5c^2$ or 10%. Therefore according to (1) the viscosity would be 1.37 cp, and corrected is 1.52 cp. Considering that hemoglobin solutions are diluted by plasma, we find that the resulting plasma/solution mixture viscosity is not significantly different from that of plasma alone when compared with the larger viscosity effect due to the presence of red blood cells.

Summation of effects from decreased blood viscosity cause the progressive increase in cardiac output and blood velocity at all levels of the circulation as hematocrit is lowered (Richardson and Guyton 1959). Between normal and about 30% hematocrit circulatory flow velocity is proportionally greater than the decrease in hematocrit, raising effective oxygen carrying capacity of the circulation. Colloidal and crystalloid volume replacement fluid viscosity is similar to that of plasma, therefore it follows that similar effects should take place when oxygen carrying blood substitutes are used if their viscosity is similar to that of plasma. When this does not follow, either the heart does not respond to improved filling or peripheral vascular resistance is increased by factors extraneous to the fluid properties of the circulating mixture.

11.2.2 Microvascular Tissue Oxygenation

The circulation is assumed to deliver oxygenated blood to the capillaries whose principal function should be that of supplying oxygen to the tissue. Current studies, however, challenge this view, since intravascular PO_2 measurements show that in many tissues oxygen tension in arterioles of about 50-μm diameter is of the order of 50 mm Hg (Duling and Berne 1973). Oxygen tension falls further in the branching arteriolar network, to the extent that capillary PO_2 is about 30 mm Hg. Correlating these PO_2 values to the oxygen saturation curve for hemoglobin (Figure 11.1) shows that tissue oxygen is supplied equally by diffusion from arterioles and capillaries. Furthermore, the arterial oxygen loss must involve the oxygen supply to the arterial vasculature, while venous oxygen is circulating surplus. Venous PO_2 is also found to increase from the capillary system onward, suggesting the presence of either arterial blood or diffusional shunts.

Measured PO_2 distribution

	PO_2, mm Hg
Systemic	100
Arterioles, 50 μm diameter	50
Terminal arterioles	30
Capillaries	20
Tissue	20
Venules, 50 μm diameter	20
Venules, 50 μm diameter	22

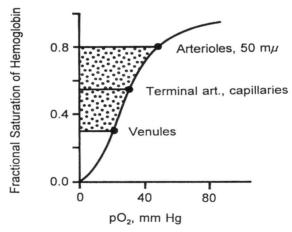

Figure 11.1 Distribution of intravascular PO_2 in the normal circulation. Direct measurements using the Pd-porphyrin technique in the awake hamster window chamber preparation. Correlation with the blood oxygen saturation curve shows that the microcirculation only delivers about half the oxygen in blood, and capillaries distribute about one-quarter of the total oxygen available.

According to Figure 11.1, the contribution to tissue oxygenation by the arterioles should be significant. We should note that arteries, a similar tissue mass, consume about 25% of the oxygen available in blood, suggesting that arterioles may do the same. Studies show that arteriolar oxygenation demonstrates a deficit, whereby the oxygen exit is greater to what is delivered to the tissue (Popel, Pittman and Ellsworth 1989). This suggests the existence of an oxygen sink at the level of the arteriolar tissue, or the existence of oxygen shunts which by diffusion or convection cause oxygen to bypass the microcirculation and enter directly the venular return, a fact in part shown by the increase of venular PO_2.

The anatomical arteriolar venular juxtaposition in a counter current configuration provides a diffusional mechanism for microvascular oxygen shunting. It is well established that arterioles, venules, lymphatics and nerves travel in a sort of communications highway through the tissue at the level of microvessels with diameters of 100 μm. Consequently, precapillary diffusional shunting is present (Mirhashemi *et al.* 1987a). A second type of shunting must also be present although not conclusively demonstrated. Capillary hematocrit is invariably about half of the systemic value. This has been explained by invoking the presence of a deformable protein layer at the capillary surface or the presence of hydrodynamic red blood cell velocity augmentation effects. Neither of these mechanism has been demonstrated conclusively, and the existence of high-hematocrit thoroughfare channels that also shunt leukocytes between arterioles and venules (Ley *et al.* 1989) may be a more realistic explanation.

11.2.3 Functional Capillary Density

Capillary oxygen delivery is dominated by the large surface area of this vascular compartment and low oxygen gradients. Their small intrinsic oxygen carrying capacity and low intravascular PO_2 renders each capillary unique in supplying a localized tissue volume, unless this is also within the diffusion field of an arteriole. Consequently, there is a portion of the tissue where capillary flow cessation, *i.e.*, the decrease of functional capillary density, is deleterious and possibly a fatal event.

The mechanism behind changes in number of capillaries with red blood cell flow has not been demonstrated. Physical considerations suggest that this may result from mechanical events comprising: 1) Capillary lumen narrowing beyond where capillary pressure can provide the energy needed for red blood cell deformation, 2) capillary luminal obstruction by leukocytes, microthrombi, rigid red blood cells, and 3) hydrodynamic effects at capillary bifurcations, which direct red blood cells to the stream with the greater flow and flow velocity. Capillary diameter variability underlies these scenarios, although it is generally assumed that capillary lumen is mostly invariable and independent of transmural pressure. This perception stems from the conclusion that capillaries are rigid, having elastic properties of the medium in which they are imbedded (Fung, Zweifach and Intaglietta 1966). It is similarly assumed that they are not contractile.

A corollary to present perceptions of capillary mechanics is that if perfusion pressure decreases, flow rate decreases throughout the microvascular network, and unless shear rate falls below the yield value for blood (if this obtains in capillary flow) capillary flow should not stop, i.e., there is no change in functional capillary density associated with lowered perfusion pressure.

Studies in the skeletal microcirculation (Lindbom and Arfors 1985) show that functional capillary density changes reversibly when perfusion pressure varies. The same study shows that functional capillary density is lowered when local tissue oxygen tension is artificially increased. Pressure flow studies in isolated organs show that flow hindrance increases as perfusion pressure decreases, a behavior mostly attributed to shear dependance of blood viscosity and diameter changes in the distensible segments of the vasculature. Decreased functional capillary density has been observed in low-flow conditions associated with ischemia reperfusion injury (Menger, Steiner and Messmer 1992). This information supports the contention that capillary perfusion pressure is the primary factor in determining the extent of functional capillary density. Lowered perfusion pressure, coupled to endothelial dysfunction, gives rise to the potential for causing pathological capillary flow hindrance as three mechanisms converge to promote this phenomenon, namely tissue edema, endothelial edema and elastic recoil.

11.3 Hemodilution with Colloids

11.3.1 Macro and Microhemodynamics

Hemodilution with colloids is implemented with macromolecules such as dextran, albumin and starch, whose molecular weight is such that intravascular retention is assured. Solutions of dextran with 70,000 molecular weight (Dextran 70) are a convenient colloidal fluid for experimental studies which is also used clinically.

Direct *in vivo* studies in tissues as diverse as skeletal muscle (Mirhashemi *et al.* 1987b, Tsai, Arfors and Intaglietta 1991), skin (Mirhashemi *et al.* 1988) and mesentery (Tigno and Henrich 1986, Lipowsky and Firrel 1986) show that for normal hematocrit, capillary hematocrit is about half of the systemic value. An important feature of capillary hematocrit is that it does not appear to fall in the same proportion as systemic hematocrit during hemodilution. Flow velocity increases monotonically in all tissues in proportion to the increase in cardiac output, and the combination of maintenance of hematocrit and increased capillary flow velocity maintains oxygen delivery capacity of the microcirculation up to systemic hematocrit reductions of the order of 75%, i.e., reductions of the circulating red blood cells to the extent that

1/4 of the original number remains. Figure 11.2 summarizes these results.

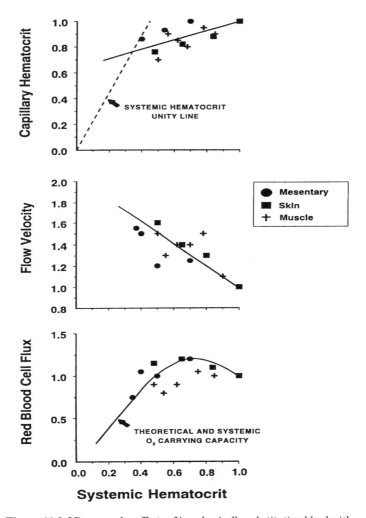

Figure 11.2 Microvascular effects of isovolemically substituting blood with 6% dextran 70,000 MW in different tissues. Mesentery: Tigno and Henrich 1986, Lipowsky and Firrell 1986; Skeletal muscle: Mirhashemi *et al.* 1987b, Tsai, Arfors and Intaglietta 1991; Awake hamster skin fold: Mirhashemi *et al.* 1988. Capillary hematocrit is maintained until reaching the line describing one to one changes of systemic and microcirculatory hematocrit (systemic hematocrit unity line). Animals generally do not survive hemodilution beyond this point. Theoretical and systemic oxygen carrying capacity line is from the study of Messmer *et al.* 1982. All values are normalized relative to the control condition where the preparation has undiluted normal blood.

11.3.2 Oxygen Distribution

Functional capillary density is maintained with ±10% of control through-out the range of hematocrit reported in these studies, which is the normal variability for this parameter. Most studies report a tendency for arterioles to vasoconstrict when hematocrit is reduced by about 30%, which is the range in which the oxygen carrying capacity of the microcirculation is maximal. This phenomenon should be due to metabolic autoregulation, a fact supported by preliminary studies on oxygen distribution during dextran 70 hemodilution from our laboratory that show that both intra- and extravascular oxygen tensions are increased relative to control in this range of hemodilution.

Studies in tissue oxygenation with multiwire Clark type electrodes show that when red blood cell mass is decreased by about 30 - 40%, tissue oxygenation is improved as the histogram of tissue oxygen distribution is shifted to the right (Messmer et al. 1982). This is the consequence of the maintained oxygen carrying capacity of both arterioles and capillaries, a decreased oxygen diffusional loss due to the higher blood flow velocity and the lack of biological reactions at the level of the endothelium due to the presence of dextran 70, resulting in the maintenance of a basal vessel wall metabolism.

Animals do not survive when hematocrit falls below about 20 - 22% in our experimental model of the awake hamster, without assisted respiration. This range of hematocrit corresponds to the intersection of capillary hematocrit with the unity hematocrit line (Figure 11.2). We presume that in these conditions the mechanism that strives to maintain hematocrit constant is no longer operative. Furthermore, at these low hematocrits, the heart is probably unable to maintain increased cardiac output. These two factors would appear to cause a non-survivable fall in oxygen carrying capacity when the red blood cell mass is reduced to 1/5 of the normal value.

11.4 Hemodilution with Crystalloids

When Ringer's lactate or normal saline is used as a plasma expander or dilutent, we find the same effects as with colloids, with the difference that the lack of oncotically active material hinders fluid absorption into the blood compartment causing fluid to extravasate. This phenomenon is readily corrected by the administration of additional fluid, being generally accepted that volume restitution with crystalloids requires the administration of about 3 times the blood volume to be replaced. The relative advantages of crystalloid vs. colloid fluid volume replacement have been extensively argued (Moss et al. 1981).

From the perspective of events in the microcirculation, there appears to be no difference between the two methods. In particular, edema that accompanies crystalloid hemodilution does not appear to affect functional capillary density or tissue function in most organs, with the exception of the lung where there is some evidence that crystalloid volume maintenance exhibits a slight increase in the incidence of adult respiratory distress syndrome.

11.5 Microvascular Effects of αα-Hemoglobin

11.5.1 Microvascular Hemodynamics

Our studies have centered on the microvascular effects of substituting blood with isooncotic cross-linked αα-hemoglobin (LAIR, San Francisco) delivered in an isovolumetric exchange. These conditions allow direct comparison with similar experiments done using dextran 70. Our experimental model is the hamster skin fold preparation, which allows observations of the microcirculation for prolonged periods and without anesthesia (Papenfuss *et al.* 1979, Endrich *et al.* 1980).

Our principal findings are that functional capillary density is compromised, falling to about 60% of control when the hematocrit is reduced by about 40%. Red blood cell capillary flow velocity is increased with no statistically significant difference with the effects found with dextran up to a decrease in hematocrit of about 50% (*i.e.*, loss of half of the red blood cell mass). When these factors are related to calculate the effective oxygen carrying capacity of the microcirculation, defined as the product of oxygen carrying capacity (red blood cell plus hemoglobin) x capillary flow velocity x functional capillary density, we find that the capacity for oxygen delivery of the hemoglobin blood mixture down to a hematocrit half of the control value is identical to that obtained with dextran 70 hemodilution (Figure 11.3).

These results indicate that this type of hemoglobin does not appear to provide a measurable benefit when compared to 50% hemodilution with non-oxygen carrying substitutes. This result may be due to either a biological effect of the material, such as scavenging of nitric oxide (Moncada *et al.* 1989), leading to vasoconstriction, or that the blood hemoglobin mixture carries too much oxygen, thus eliciting a metabolic autoregulatory effect. Both mechanisms may be operational, however, the oxygen effect may be prevalent.

Further hemodilution with αα-hemoglobin shows the benefit of increased intrinsic oxygen carrying capacity, and when 3/4 of the red blood cell mass is replaced the oxygen carrying capacity of the microcirculation is double that obtained with the same level of hemodilution with a non-oxygen carrying material. As may be expected at this level of hemodilution, the autoregulatory responses are reversed, and microvascular diameters return to control values. Hemodilution to this level with αα-hemoglobin is tolerated by the animals in our experimental model, this not being the case when dextran 70 is used.

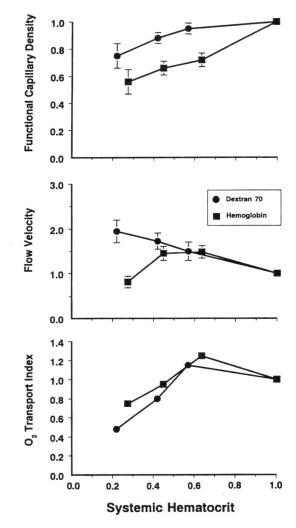

Figure 11.3 Microvascular effects of isovolemically substituting blood with 15% solution of αα-hemoglobin in the awake hamster window preparation. Comparison to identical experiments with 6% dextran 70,000 MW. Flow transport index is calculated as the product of capillary flow velocity times capillary hemoglobin concentration (red blood cells plus free hemoglobin) times functional capillary density. All values are normalized relative to the control condition where the preparation has undiluted normal blood.

11.5.2 Oxygen Transport

Microvascular PO_2 in the subcutaneous tissue is determined with the phosphorescence decay technique based on the oxygen-dependent quenching of phosphorescence emitted by metallo-porphyrins bound to albumin [Torres and Intaglietta 1993, Torres *et al.* 1994). In this process phosphorescence lifetime decreases in proportion to the local oxygen tension. The relationship between phosphorescence lifetime τ and PO_2 is described by the Stern-Volmer equation (Vanderkooi *et al.* 1987, Wilson 1993). The porphyrin compound remains initially within the blood vessels and diffuses into the interstitium after about one hour, allowing the determination of intra-arterial, intravenous and tissue PO_2. The technique is independent of absolute light intensities and porphyrin concentration.

Specifically, palladium-meso-tetra (4-carboxyphenyl) porphyrin (Porphyrin Products, Inc., Logan, UT USA) bound to bovine serum albumin and diluted in saline (sodium chloride 0.9%, Elkins-Sinn, Inc., Cherry Hill, NJ USA) to a concentration of 15 mg/ml is injected intravenously in a dosage of 30 mg/kg body weight. Current versions of this instrumentation developed in our laboratory allow the measurement of PO_2 in rectangular areas of 10 x 100 μm.

Oxygen measurements carried out under control conditions are shown in Figure 11.1. Measurements were repeated during conditions of high hemodilution with αα-hemoglobin (hematocrit 22%) with the results shown in Table 11.1. The finding that maintenance of arteriolar PO_2 in vessels of about 50 μm diameter led to a significant decrease in tissue PO_2 suggested that this parameter be investigated in the immediate vicinity of the arteriolar wall.

Measuring intravascular PO_2 and PO_2 next to the blood/tissue interface provides information on the rate of oxygen consumption by the vascular wall, which is directly related to microvascular wall metabolism involved in mechanical work for the regulation of blood flow and on going biochemical synthesis (renin, prostaglandins, collagen, conversion of angiotensin I to II, degradation of bradykinin and prostaglandins, and the clearance of lipids and lipoprotein). Present accuracy and resolution of the technology allows measurements in vessels of the order of 50-μm diameter where the gradient in the wall is measured in a tissue shell about 7.5-μm thick. The measured gradient is directly proportional to the rate of oxygen utilization by the tissue where the gradient is measured.

Our results show that the presence of large amounts of αα-hemoglobin generates biological activity at the level of the microvascular wall, leading to the increase of oxygen consumption by the vessel wall. The mechanism responsible for this effect is not presently known, however, the tendency for arteriolar microvascular PO_2 to increase, the increase in oxygen carrying capacity leading to metabolic autoregulation, and the biological activity associated with the release of heme and its introduction into the endothelial cell environment are mechanisms that may be directly linked to the observed increase in wall oxygen metabolism. The lowering of venular and tissue PO_2 are also indicative of the development of an extra-oxygen sink, which appears located at the microvascular wall. Early studies on the effect of introducing large amounts of free hemoglobin in the circulation report the doubling of whole body metabolism (Kendrick 1964).

Table 11.1 Oxygen tension distribution: Control (normal blood) vs. $\alpha\alpha$-hemoglobin (15 g/100 ml hemodilution to 70% reduction of hematocrit, 50 μm arterioles.

Compartment	Control, mm Hg	$\alpha\alpha$-Hb, mm Hg
Intravascular, arterioles[1]	52.7	56.7
Extravascular, arterioles	34.9	26.6*
Wall PO_2 gradient[2]	17.8	26.6*
Tissue PO_2[3]	20.9	10.8*
Intravascular, venules > 50μm	33.2	18.8*

[1]Arteriolar diameters are 50 μm ± 20%; [2]Wall PO_2 gradient develops over a distance of approximately 7.5 μm; [3]Tissue PO_2 is measured in areas comprising capillaries but void of arterioles and venules.
*Statistically significantly different from control, p > 0.05.

The present findings only provide overall evidence for the existence of potentially significant oxygen consumption at the microvascular wall. It is not possible to discern whether endothelium or smooth muscle are the major recipients of this oxygen because of the spatial resolution of the present technology. Endothelium is capable of a metabolic activity which can be 100 fold that of other cell species (Bruttig and Joyner 1983). Furthermore, in arterioles, smooth muscle mass is significantly greater than that of the endothelial component. This suggests that in the larger arterioles, oxygen use is mainly due to smooth muscle metabolism, with this prevalence shifting to the endothelium in the capillaries, where their large surface area causes them to become a significant oxygen sink.

It may be argued that the decrease in functional capillary density causes tissue PO_2 to be lower, therefore inducing the gradient at the arteriolar wall to become steeper. This is not the case in the present situation since the relationship between vessel wall gradient and metabolism obtains from the solution of the diffusion equation in cylindrical coordinates. This model for oxygen exchange between microvessels and tissue does not make any assumption about PO_2 at the boundary. The result that the wall PO_2 gradient is directly related to metabolism is solely a function of the impermeable boundary condition between diffusion fields of neighboring vessels.

11.6 The Distribution of Oxygen in Hemodilution

11.6.1 Oxygen Distribution in Leaky Vessels

Outward oxygen diffusion from arterial vessels is present throughout the circulation since there is no specific barrier to its passage across the vessel wall other than the exit rate limitation imposed by diffusion (Mirhashemi *et al.* 1987a, Ellsworth and Pittman 1990). In this context oxygen transport through the circulation is fundamentally different from the transport of water through a system of impermeable pipes. In the later mass balance at any segment shows that entrance and exit flows are identical and independent of flow rate. In permeable systems mass balance includes the exiting flow fraction due to permeation (diffusion) and the relative difference between input and output is a function of flow rate.

Experimental findings from our laboratory support this analysis. Direct measurements of microvessel intravascular PO_2 correlated to locally measured blood flow velocity (Intaglietta, Silverman and Tompkins 1975) show that changes in blood $\Delta PO_{2,\text{blood}}$ and changes in flow velocity Δv can be empirically correlated by the expression:

$$\Delta PO_{2,\text{blood}} = 2.6\ \Delta v \text{ mm Hg; } (v = \text{mm/sec); } p < 0.001$$

The significance of this expression is that change in cardiac output affects proportionately all segments of the vascular tree, since flow velocity is a direct function of the level of branching. As an example, doubling cardiac output doubles flow velocity in capillaries and arterioles equally, however, due to branching, capillary flow velocity is 0.5 mm/sec, leading to an increase of PO_2 of 1.3 mm Hg, while in 50-µm arterioles with flow velocities of about 5 mm/sec, the increase in PO_2 is 13 mm Hg.

11.6.2 Model Analysis of Oxygen Transport and Diffusion in the Circulation

The relationship between blood flow velocity, oxygen carrying capacity and rate of oxygen leakage is given by a mass balance equation for any segment of the vasculature. For any axial length dx the amount of oxygen entering the upstream cross-section at x is m_x given by:

(2) $\qquad m_x = QF_1(\%RBC, \%Hb)nPO_2$

where Q is the entering volumetric flow rate, F_1 measures the intrinsic oxygen carrying capacity of the red blood cell and/or hemoglobin mixture, and nPO_2 is the linearized functional relationship between the red blood cell (+ hemoglobin) oxygen carrying component and the PO_2 which is usually given by the oxygen saturation curve. The oxygen content after the blood mixture traverses an axial distance dx will be decreased by the rate of oxygen exit across the vessel wall. A simplified analysis equates this oxygen leakage by diffusion m_{dif} to a permeability coefficient p and the difference in PO_2 between blood and tissue. A further simplification can be obtained by assuming that tissue PO_2 is zero. Given these premises, the change in blood oxygen content as a function of distance is given by the differential relationship:

$$(3) \qquad \frac{dm_x}{dx} = QF_1 n \frac{dPO_2}{dx} = pPO_2$$

This differential equation is in terms of PO_2 rather than oxygen concentration. Oxygen diffusion is driven by the oxygen partial pressure, which in a gas is proportional to oxygen concentration. In blood, oxygen concentration is related to PO_2 through the oxygen saturation curve, which describes the oxygen held in chemical binding. To understand why oxygen mass balance in this system should be expressed as a function of partial pressures rather than concentrations (molecule per unit-volume) consider the blood/vessel wall interface. Within an infinitesimal distance in either direction from the interface, PO_2 is identical, however, oxygen concentration changes abruptly. In the tissue, oxygen diffusion is driven by the PO_2 gradient which is not related to the amount of oxygen in blood. (Similarly, the exit rate of water from a lake through a river is determined by the difference in hydrostatic head (pressure) between lake and river exit and not the size of the lake.)

Integration of the local mass balance equation yields:

$$(4) \qquad PO_2 = K_1 e^{-\frac{K_2}{C_{Hb^v}}}$$

where K_1 is a reference oxygen partial pressure, K_2 is a constant incorporating information on the permeability coefficient, the distance between reference partial pressure K_1 and the point at which PO_2 is evaluated and the linearized relationship between hemoglobin oxygen content and PO_2, and C_{Hb} is the total amount of hemoglobin comprising red blood cells and hemoglobin in solution. The values of K_1 and K_2 can be obtained directly from the distribution of oxygen in the circulation. At normal hematocrit ($C_{Hb} = 1.0$, $v = 1.0$), arterial oxygen is 100 mm Hg, and according to Figure 11.1, this value drops to 50 mm Hg in arterioles of 50 mμ allowing one to solve for K_2 from:

$$(5) \qquad 50\,(mm\,Hg) = 100 e^{\frac{-K_2}{1.0}}\,(mm\,Hg)$$

The constants for PO_2 fall to the beginning of the capillaries or terminal arterioles, and the capillary exit can be similarly evaluated. These constants are the numerical expression of the physical and anatomical determinants of the relationship between flow velocity and diffusion. In combination with the actual and intrinsic oxygen carrying capacity of the flowing mixture they can be used to predict the oxygen fall along key points of the circulation for different hematocrits and colloidal concentrations if we know or predict the velocity response of the circulation.

Equation 4 shows the critical role played by the velocity of blood in lowering the diffusional oxygen losses. Since increased velocity is usually due to decreased intrinsic oxygen carrying capacity of blood, *i.e.*, lower hematocrit, improved blood oxygenation will only manifest itself when the product $C_{Hb}v$ is greater than normal (1.0). As a corollary, any decrease of intrinsic blood oxygen carrying capacity not accompanied by increased flow velocities augments diffusional oxygen losses. Conversely, at constant intrinsic oxygen carrying capacity, the circulation delivers proportionally more oxygen at higher flow rates, since the diffusional loss is constant.

11.6.3 Relationship to Experimental Findings

Utilizing available microvascular data (Figures 11.1, 11.2 and 11.3) we can evaluate oxygen distribution in the microcirculation for different blood compositions as shown in Table 11.2.

Table 11.2 Oxygen distribution in the microcirculation following changes of intrinsic oxygen carrying capacity affecting blood flow velocity.

Colloid	Control	Dx70, 6%	Hb, 15%	Hb, 15%[*]
Total hemoglobin[1]	1.0	0.7	1.0	1.0
Hematocrit	1.0	0.7	0.7	0.3
Velocity	1.0	1.6[2]	1.5	1.8
PO_2, mm Hg	PO_2, K_2[3]		PO_2	
50-μm arterioles	50, -0.69	53	57	68
Terminal arterioles	30, -0.51	33	45	51
Capillary exit	20, -0.40	23	34	41

Data from Figures 11.1, 11.2 and 11. 3. Oxygen tension calculated from Equation 3. [1]Red blood cells plus αα-hemoglobin in solution referred to control values of 15 g/100 ml. [2]Derived from measured red blood cell flux. [*]Calculated assuming that hemoglobin would have the same hemodynamic effect as dextran 70. [3]K_2 is calculated from the oxygen distribution at control.

The results of Table 11.2 are important because they show that increased oxygen carrying capacity accompanied by lowered viscosity causes a significant increase in the oxygen levels in the arteriolar system. Comparing hematocrit reductions of 30% from control obtained with dextran 70 and $\alpha\alpha$-hemoglobin we see that while dextran only increases PO_2 by about 3 mm Hg, the use of hemoglobin and corresponding maintenance of intrinsic oxygen carrying capacity at normal levels (equivalent to that of normal blood) increase arteriolar and terminal arteriolar PO_2 by 13 and 15 mm Hg, respectively. Such a large increase in PO_2 must per force elicit an autoregulatory response, since a primary function of these vessels is oxygen regulation, a process driven by PO_2 signaling. Regulation manifests in vasoconstriction, which reduces flow, and therefore, according to Equation 3, increases oxygen exit rate from the vessels, causing local PO_2 levels to return to normal.

The presence of the fore-mentioned oxygen regulatory process is illustrated by calculating oxygen distribution in the arterioles in conditions of extreme hemodilution with $\alpha\alpha$-hemoglobin under the assumption that the hemodynamic effect would be identical to that obtained with dextran 70. In this hypothetical situation the oxygen level would be increased by about 20 mm Hg, which is clearly a non-physiological situation, leading to a strong vasoconstrictor response. The presence of this mechanism is shown by large-arteriole oxygen tension rising only to 57 mm Hg (Table 11.1 and 11.2). Calculating from Equation 3, the velocity necessary to obtain this autoregulatory response (the flow velocity that in this condition of hemodilution would lead to a large-arteriole PO_2 of 56.7 mm Hg) leads to the value of 1.2, which compares to the value of 1.0 in Figure 11.3 (0.3 systemic hematocrit).

A decrease in flow velocity from 1.8 (Htc 0.3) present with dextran 70 hemodilution to 1.0 requires that the pressure gradient be reduced in the same proportion, leading to capillary pressure reduction of the order of 45%. Although the distribution of microvascular pressures has been extensively studied since the development of the servo-nulling micropressure technique by our group (Intaglietta, Pawula and Tompkins 1970), most studies have relied on anesthetized preparations and exposed tissue, and the actual values in the normal tissue are only approximately known. Estimates of capillary pressure for anesthetist mammals set this number at 30 ± 3 mm Hg (Lipowsky 1987). Therefore, a reduction of 45% will significantly reduce functional capillary density according to the findings of Lindbom and Arfors (1985), fully corroborating our results (Figure 11.3).

11.7 Conclusions

The decrease of blood viscosity while maintaining intrinsic oxygen carrying capacity is a relatively unexplored condition in terms of hemodynamic transport phenomena. The present theoretical and experimental findings lead to the following microvascular perturbation:

1) Lowering blood viscosity increases arteriolar blood oxygen content to the extent that autoregulatory mechanisms are engaged.

2) Engagement of arteriolar oxygen metabolic autoregulation lowers functional capillary density due to vasoconstriction and lowered capillary pressure.

3) Relating the PO_2 distribution to the oxygen dissociation curve for hemoglobin shows that the amount of oxygen delivered in the microcirculation is significantly decreased during hemoglobin hemodilution, even in the absence of lowered functional capillary density.

4) Experimental data on the development of an increased wall oxygen gradient (Table 11.1) shows that this process further contributes to the impairment of tissue oxygenation with hemoglobin solutions.

Items 1 through 3 are a direct consequence of altered transport properties and may be modified by tailoring the properties of the hemoglobin solution to diminish PO_2 redistribution. Left-shifting the oxygen dissociation curve would accomplish the objective. Increasing the viscosity of the hemoglobin diluted blood would also attain the same objective and may be desirable in situations where increasing cardiac output is not indicated. Implementation of either strategy or their combination may lower the wall oxygen gradient since this is in part due to increased mechanical and metabolic work required by the blood vessel in producing vasoconstriction. Maintenance of normal arteriolar PO_2 distribution in the presence of high intrinsic microcirculatory oxygen capacity is necessary to avoid an oxygenation paradox, whereby tissue becomes ischemic in the presence of adequate oxygenation and flow.

11.8 Acknowledgments

This research was supported in part by the National Heart, Lung and Blood Institute of the National Institutes of Health (P01 HL48018).

11.9 References

Bruttig, S.P., and W.L. Joyner. Metabolic characteristics of cells cultured from umbilical blood vessels: Comparison with 3T3 fibroblasts. *J. Cell. Physiol.* 116: 173-180, 1983.

Duling, B.R., and R.M. Berne. Longitudinal gradients of perivascular oxygen tension. *Circ. Res.* 27: 669-678, 1973.

Einstein, A. A new determination of molecular dimensions. In *The Theory of Brownian Motion*. Chapter 3, Dover Publications, Inc., 1956.

Ellsworth, M.L., and R.N. Pittman. Arterioles supply oxygen to capillaries by diffusion as well as by convection. *Am. J. Physiol.* 258: H1240-H1243, 1990.

Endrich, B., K. Asaishi, A. Götz, and K. Messmer. Technical report: A new chamber technique for microvascular studies in unanaesthetized hamsters. *Res. Exp. Med.* 177: 125-134, 1980.

Fung, Y.C., B.W. Zweifach, and M. Intaglietta. Elastic environment of the capillary bed. *Circ. Res.* 19: 441-461, 1966.

Kendrick, D.B. *Blood program in World War II.* Washington D.C.: Office of the Surgeon General, 1964.

Intaglietta, M., R.F. Pawula, and W.R. Tompkins. Pressure measurements in the mammalian microvasculature. *Microvasc. Res.* 2: 212-220, 1970.

Intaglietta, M., N.R. Silverman, and W.R. Tompkins. Capillary flow velocity measurements *in vivo* and *in situ* by television methods. *Microvasc. Res.* 10: 165-179, 1975.

Intaglietta, M. Microcirculatory effects of hemodilution: background and analysis. In *The Role of Hemodilution in Optimal Patient Care* (R.F. Tuma, J.V. White and K. Messmer, eds.) München: W. Zuckschwerdt Verlag, 1989, pp. 21-41.

Ley, K., J.-U. Meyer, M. Intaglietta, and K.-E. Arfors. Shunting of leukocytes in rabbit tenuissimus muscle. *Am. J. Physiol.* 34: H85-H93, 1989.

Lindbom, L., and K.-E. Arfors. Mechanism and site of control for variation in the number of perfused capillaries in skeletal muscle. *Int. J. Microcirc.: Clin. Exp.* 4: 121-127, 1985.

Lipowsky, H.H., S. Usami, and S. Chien. *In vivo* measurement of apparent viscosity and microvessel hematocrit in the mesentery of the cat. *Microvasc. Res.* 19:297-319, 1980.

Lipowsky, H.H., and J.L. Firrel. Microvascular hemodynamics during systemic hemodilution and hemoconcentration. *Am. J. Physiol.* 250: H908-H922, 1986.

Lipowsky, H.H. Mechanics of blood flow in the microcirculation. In *Handbook of Bioengineering* (R. Skalak and S. Chien, eds.) New York: McGraw Hill Book Co., Chapter 18, 1987.

Menger, M.D., D. Steiner, and K. Messmer. Microvascular ischemia-reperfusion injury in striated muscle: significance of "no-reflow". *Am. J. Physiol.* 263: H1892-H1900, 1992.

Messmer, K., L. Sunder-Plassman, H.V. Hessler, and B. Henrich. Hemodilution in peripheral occlusive disease: a hematological approach. *Clin. Hemorheol.* 2: 721-731, 1982.

Mirhashemi, S., S. Ertefai, K. Messmer, and M. Intaglietta. Model analysis of the enhancement of tissue oxygenation by hemodilution due to increased microvascular flow velocity. *Microvasc. Res.* 34: 290-301, 1987a.

Mirhashemi, S., K. Messmer, K.-E. Arfors, and M. Intaglietta. Microcirculatory effects of normovolemic hemodilution in skeletal muscle. *Int. J. Microcirc.: Clin. Exp.* 6: 359-370, 1987b.

Mirhashemi, S., G.A. Breit, R.M. Chávez, and M. Intaglietta. Effects of hemodilution on skin microcirculation. *Am. J. Physiol.* 254: H411-H416, 1988.

Moncada, S., R.M.J. Palmer, and E.A. Higgins. Biosynthesis of nitric oxide from L-arginine. *Biochem. Pharmacol.* 38: 1709-1715, 1989.

Moss, G.S., R.J. Lowe, J. Jilek, and H.A. Lavine. Colloid or crystalloid in the resuscitation of hemorrhagic shock: a controlled clinical trial. *Surgery* 89: 434-438, 1981.

Papenfuss, H.D., J.F. Gross, M. Intaglietta, and F.A. Treese. A transparent access chamber for the rat dorsal skin fold. *Microvasc. Res.* 18: 311-318, 1979.

Popel, A.S., R.N. Pittman, and M.L. Ellsworth. Rate of oxygen loss from arterioles is an order of magnitude higher than expected. *Am. J. Physiol.* 256: H921-H924, 1989.

Quemada, D. Rheology of concentrated dispersed systems: III. General features of the proposed non-Newtonian model: comparison with experimental data. *Rheol. Acta* 17: 643-653, 1978.

Richardson, T.Q, and A.C. Guyton. Effects of polycythemia and anemia on cardiac output and other circulatory factors. *Am. J. Physiol.* 197: 1167-1170, 1959.

Tigno, X.T., and H. Henrich. Flow characteristics of the microcirculation following intentional hemodilution. *Acta Med. Phil.* 22: 5-12, 1986.

Torres, Filho, I.P., and M. Intaglietta. Microvessel PO_2 measurements by phosphorescence decay method. *Am. J. Physiol.* 265: H1434-H1438, 1993.

Torres, Filho, I.P., Y. Fan, M. Intaglietta, and R.K. Jain. Noninvasive measurement of microvascular and interstitial oxygen profiles in a human tumor in SCID mice. *Proc. Natl. Acad Sci. USA* 91: 2081-2085, 1994.

Tsai, A.G., K.-E. Arfors, and M. Intaglietta. Spatial distribution of red blood cells in individual skeletal muscle capillaries during extreme hemodilution. *Int. J. Microcirc.: Clin. Exp.* 10: 317-334, 1991.

Vanderkooi, J.M., G. Maniara, T.J. Green, and D.F. Wilson. An optical method for measurement of dioxygen concentration based upon quenching of phosphorescence. *J. Biol. Chem.* 262: 5476-5482, 1987.

Wilson, D.F. Measuring oxygen using oxygen dependent quenching of phosphorescence: a status report. *Adv. Exp. Med. Biol.* 333: 225-232, 1993.

Chapter 12

Oxygen Delivery Regulation: Implications For Blood Substitutes

*Paul C. Johnson, Ph.D., Keith Richmond, Ross D. Shonat, Ph.D., Andreas Toth, Mihlos Pal, M.D., Marc E. Tischler, and Ronald M. Lynch

Department of Physiology, University of Arizona College of Medicine, Tucson, Arizona 85724

Department of Bioengineering, University of California, San Diego, La Jolla, California 92093-0412

ABSTRACT

All of the variables regulating the delivery of oxygen to specific tissues have not been identified. However, O_2 capacity, blood flow, and vessel diameter, are coordinated in concert to optimize tissue oxygenation. Experimental models studied to date indicate that the set points for muscle blood flow are well above those needed to maintain oxidative metabolism. Thus, additional mechanisms must exist. A possible point of control is the supply of O_2 to arterioles which can regulate capillary blood flow. This hypothesis has important implications for the development of red cell substitutes.

12.1 Introduction

Blood flow regulation and oxygen delivery to the tissues are closely intertwined. On the one hand, blood flow to many organs appears to be closely regulated such that the parenchymal cells receive an adequate supply of oxygen to satisfy the requirements of oxidative metabolism under a wide variety of circumstances. On the other hand, blood flow regulatory mechanisms also appear to prevent an over-abundance of oxygen delivery to the tissues.

Blood Substitutes: Physiological Basis of Efficacy
Winslow et al., Editors
© Birkhäuser Boston 1995

12.2 Autoregulation: General Concepts

A classical example of how blood flow is determined by demand is the proportionate increase in blood flow in working skeletal muscle with increase in oxygen consumption (Sparks 1978) as shown in Figure 12.1. Note also the reduction in venous oxygen levels, especially at lower work rates. Another phenomenon thought to be due in part to the same

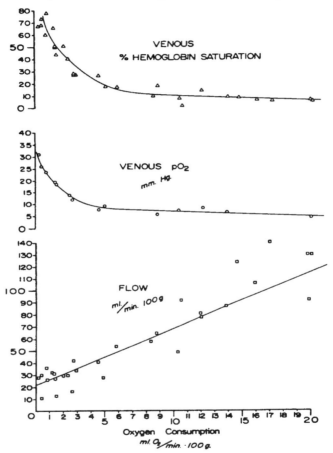

Figure 12.1 Relationships among oxygen consumption, blood flow and venous PO_2 of denervated skeletal muscle in the dog hind limb. The oxygen consumption was varied by motor nerve stimulation rates of 0.25 to 8 Hz. From Sparks (1978) by permission.

mechanism is autoregulation of blood flow, that is, the tendency for blood flow to remain constant despite changes in arterial perfusion pressure (Johnson 1986). Autoregulation in skeletal muscle is enhanced in working muscle, perhaps because vascular tone is more closely attuned to metabolic needs than in resting muscle (Stainsby 1962, Johnson 1986).

Based on observations such as these, physiologists have sought for many years to determine the exact nature of the linkage between blood supply and oxygen demand. The working hypothesis of these studies has been that when the oxygen supply to the tissues is inadequate a signal in the form of a chemical vasodilator is transmitted from the parenchymal cells to the resistance vessels, reducing the tone of these vessels and allowing blood flow to increase to meet metabolic needs. Most commonly, it has been thought that this chemical signal arises when tissue oxygen tension falls below critical levels and there is a shift from oxidative metabolism to anaerobic glycolysis (Schubert, Whalen and Nair 1978).

It must be admitted that after more than 100 years of study, we still cannot, with a high degree of confidence, identify the chemical mediator or mediators responsible for functional hyperemia of working muscle or evidence for involvement of such mediators in autoregulation of blood flow. There are, however, several candidates such as adenosine, H^+ ion, and potassium for which there is some evidence in certain organs (Sparks 1978, Johnson 1986).

While there is abundant evidence that blood flow is regulated in such a manner as to provide the oxygen necessary to support oxidative metabolism, it is also evident that blood flow to tissues such as skeletal muscle is regulated in such a manner as to prevent an over-abundance of oxygen in the tissues.

A very simple experiment that illustrates this principle is the suffusion of an oxygen-rich solution over an exposed microcirculatory bed in a muscle such as the cremaster muscle of the rat or hamster. When this is done, the arterioles constrict and red cell velocity in the arterioles falls as shown in Figure 12.2 (Prewitt and Johnson 1976). This mechanism is so effective that the fall in blood flow can completely offset the increased oxygen delivery from the suffusing solution, and there is little net change in tissue oxygen tension, which in skeletal muscle is typically in the range of 20 mm Hg (Duling 1978, Prewitt and Johnson 1976).

Given the fact that tissue oxygen tension appears to be held within a certain narrow range in skeletal muscle, it is possible that there are separate mechanisms coupling tissue oxygen demands to blood flow; one causing vasodilation when blood flow falls too low, and another causing vasoconstriction when flow rises too high. It is also possible that there is a single mechanism with a set point around the normal tissue PO_2 level of about 20 mm Hg. In the latter case, it is supposed that other mechanisms such as the myogenic response and neural adrenergic mechanisms provide a basal level of vascular tone that is antagonized to different degrees by the action of the purported vasodilator metabolite or metabolites. Note that in either case, it is hypothesized that the low end of the "acceptable" range of blood flow is determined by release of vasodilator substances when tissue oxygen tension falls too low. One form of this hypothesis is that there are normally tissue areas on the borderline of hypoxia, and these areas are normally producing just sufficient quantities of vasodilator metabolites to maintain adequate tissue oxygenation overall (Schubert, Whalen and Nair 1978). Such areas for example might be present near the venous end of the capillary network.

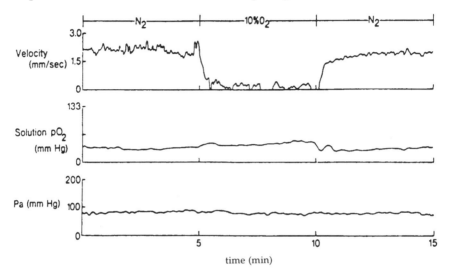

Figure 12.2 Red cell velocity in a 20 μm arteriole of rat cremaster muscle with alteration of the oxygen level of the suffusing solution. The top line indicates the gas mixture with which the suffusate was equilibrated. Solution PO_2 was measured at the surface of the muscle. Pa is arterial pressure. From Prewitt and Johnson (1976) by permission.

12.3 Experimental Studies

In our studies of this area we have set out to examine the question of the linkage between oxygen delivery and oxygen demand by determining whether blood flow is normally held at a level close to the minimum required for adequate tissue oxygenation. Specifically, we have set out to determine whether in resting muscle there are tissue areas that are either frankly hypoxic or on the verge of hypoxia.

Before describing these studies, I would like to make clear that the mechanisms being considered are relevant only to organs in which the ratio of oxygen consumption to oxygen delivery is relatively high, such as skeletal muscle, myocardium and brain. These considerations are not necessarily relevant to organs such as kidney and secretory glands where the blood supply is much higher than needed to meet the requirements of oxidative metabolism. Obviously, in those organs, factors other than those related to oxidative metabolism determine the blood flow to the tissue.

The studies I will describe are also pertinent from the standpoint of designing an adequate blood substitute. For this purpose, it is of course very important to know how much oxygen is required to provide adequate delivery to all areas of the tissue. We have chosen to address this question at the microcirculatory level by lowering blood flow and determining the point at which pyridine nucleotide fluorescence in the tissue begins to rise. We selected pyridine nucleotide because it is a coenzyme involved in the electron transport chain, and when oxygen supply becomes inadequate, there is a shift from the oxidized NAD^+ to the reduced NADH. Fortunately, NADH fluoresces while NAD^+ does not, so a rise in this signal gives an indication of the shift from aerobic to anaerobic metabolism (Chance *et al.* 1962).

12.3.1 Apparatus

The experimental arrangement used consists of a microscope system on which an anesthetized animal can be mounted on the microscope stage and a thin muscle such as the cat sartorius or rat spinotrapezius exteriorized for direct study (Toth *et al.* 1992). With this arrangement the muscle can be transilluminated with a Xenon light source to view the microcirculation with a video camera placed in the image plane of the microscope objective. A second, mercury arc source provides the excitation wavelength (366 nm) for the NADH fluorescence. The 450 nm fluorescence signal from the endogenous NADH in the muscle is collected by the objective and passes to a cooled photomultiplier tube and a photon counter and is stored in a computer. With this system we are able to monitor NADH fluorescence in a 20 µm tissue area. The small sample site enables us to monitor NADH levels in regions devoid of blood vessels. This is important since hemoglobin absorbs both the excitation wavelength and the emitted fluorescence. The field is imaged with the video camera and recorded on video tape for off-line analysis of red cell velocity in capillaries adjacent to the site of NADH fluorescence measurement.

12.3.2 Stopped-flow Experiments

Several types of studies have been performed with this system. First, we have simply stopped flow to the muscle and determined how long it takes before a shift to anaerobic metabolism occurs, as judged by the rise in NADH fluorescence signal. According to the hypothesis we are testing, there should be a number of tissue areas that are hypoxic or on the verge of hypoxia. Such areas should show a prompt rise in fluorescence when blood flow stops. On the other hand, tissues near the arterial end of the capillary network might be better supplied with oxygen and show a longer lag time before NADH rises.

A summary of our findings is shown in Figure 12.3. These data were taken from 20 μm spots near the arteriolar and venous ends of the capillary network. As flow stopped, there was typically a latent period of about 45 seconds, and NADH fluorescence then began to rise, reaching a peak level of about 55% above control after about 100 seconds. There was no significant difference between arterial and venous sites in respect to the time course or magnitude of change in NADH fluorescence. The absence of a difference in delay time at these two regions was unexpected, in as much as oxygen is unloaded as red cells travel through the capillaries and the oxygen tension at venous sites is lower than at the arterial sites.

The difference in PO_2, however, is not as great as might be expected from idealized models such as the Krogh cylinder, because blood flow in capillary networks adjacent to each other may be counter-current, and the arterial region of one network may lie adjacent to the venous region of another (Koller and Johnson 1986). The findings may also reflect the fact that venous capillaries are more numerous and have larger diameters than those at the arterial end of the network, providing a greater blood volume at the venous end and thus a larger reservoir for oxygen.

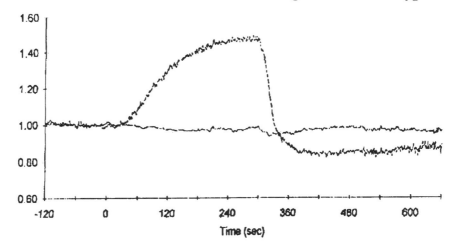

Figure 12.3 NADH fluorescence and transmittance of the excitation wavelength in the cat sartorius muscle during a five minute period of flow stoppage. Fluorescence rises after a delay period while the transmittance does not change. The latter indicates that the optical density of the tissue area from which the fluorescence arises did not change during the period of measurement. Average data from 60 tissue sites, 30 near the venular end of the capillary network and 30 near the arteriolar end. Optical signals are normalized to values during the control period before occlusion.

I should mention that there is a substantial amount of NADH present even in well-oxygenated tissue. Hypoxia shifts the ratio of $NAD^+/NADH$. Studies in the sartorius muscle with free blood flow show that the NADH content is 0.322 ± 0.014 nmol/mg protein (N=7), while in a muscle that has been ischemic for 5 minutes, the NADH level rises to 0.563 ± 0.024 nmol/mg protein (N=7). In addition, not all of the fluorescence measured at 450 nm arises from NADH; other substances such as collagen also contribute. However, it is unlikely that the fluorescence from such other sources changes during a brief period of ischemia.

12.3.3 Reduced-flow Experiments

A second type of study we have performed is to reduce rather than stop flow in the sartorius muscle and observe the degree to which flow must be reduced in order to elicit a shift from oxidative to anaerobic metabolism. This was performed by stimulating the sympathetic nerve to the muscle for a period of 3 minutes while monitoring NADH fluorescence in a localized tissue area and blood flow in the adjacent capillaries. A fall of less than 50% in red cell velocity in the adjacent capillaries failed to cause an increase in NADH. However, if the fall in red cell velocity was more than 50% and the decrease was sustained, NADH would rise. In general, we found that a decrease in blood flow of at least 50% sustained for at least 30 seconds was required to increase the NADH fluorescence signal.

This study supports the observations described above with complete flow stoppage and indicates that blood flow is normally regulated at levels well above that required to maintain oxidative metabolism in the tissue regions that would be most likely to be on the verge of hypoxia.

12.3.4 Tissue PO_2

Previous *in vitro* studies on isolated mitochondria suggest that the critical PO_2 at which the shift from aerobic to anaerobic energy metabolism occurs is well under 1 mm Hg (Longmuir 1957). To this time there have not been equivalent studies *in vivo*. It has been suggested that the critical PO_2 in whole cells might be quite different from that in washed, isolated mitochondria due to a lower affinity of the cytochrome chain for oxygen (Rosenthal *et al.* 1976).

To investigate this issue we have added to our microscope system the capability to monitor tissue PO_2 with a palladium porphyrin probe whose phosphorescence lifetime varies according to the oxygen partial pressure (Vanderkooi *et al.* 1987, Torres Filho and Intaglietta 1993). For this purpose a strobe light source was interposed above the objective to provide brief flash epi-illumination of the tissue. A photomultiplier tube collected the phosphorescence signal which was fed into a computer that determined the signal life-time and PO_2. The probe was injected intravenously into the animal and, since it is linked to albumin, first distributes in the vascular compartment and then slowly leaks out into the tissue. After an hour or so, there is sufficient probe in the tissue, probably localized in the extracellular space, to monitor PO_2 there as well as in the adjacent microcirculatory vessels. With this system we were able, at the same site, to determine both NADH and PO_2.

The aim of this study was to determine the tissue PO_2 and the PO_2 in adjacent vessels at which the shift from oxidative to anaerobic metabolism occurs. To this end we have performed studies in which, as in the first series of experiments described above, we stopped blood flow and monitored the rise in NADH fluorescence. Tissue PO_2 could not be monitored simultaneously so we have done consecutive experiments in which the decline of PO_2 during the period of flow stoppage or the rise in NADH was monitored. An example of our findings is shown in Figure 12.4. The tissue PO_2 falls in a linear fashion and then shifts abruptly to

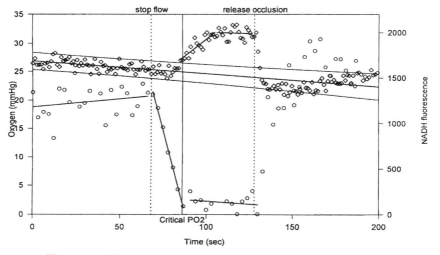

Figure 12.4 Tissue NADH and oxygen tension in a 20 μm diameter tissue site adjacent to a venule in the rat spinotrapezius muscle during a one minute period of occlusion. Upper symbols represent the NADH fluorescence and the lower symbols the oxygen tension. Also shown are the 95% confidence level for the NADH signal and the point at which this signal increases significantly during the period of occlusion. This point was considered to represent the critical PO_2 for the tissue site. Data were obtained from consecutive experiments in which NADH fluorescence and O_2 tension were separately measured at the same site.

a much more shallow slope at a value of about 2 mm Hg. At this same point NADH begins to rise. When PO_2 in an adjacent venule with about a 20 μm diameter is monitored, we find that the shift in metabolism occurs at a blood PO_2 of about 7 mm Hg.

12.3.5 Implications of Experimental Studies

From the standpoint of blood flow regulation, these studies suggest that the set point for blood flow, at least in resting cat sartorius muscle, is well above that needed to maintain oxidative metabolism. The findings do not support the hypothesis that there is an hypoxic state in some tissue regions that provides a vasodilator chemical signal from these regions. Rather it appears that the tissue PO_2 is regulated at such a level that all tissue regions are well supplied with oxygen.

The implication of our findings is that blood flow is controlled by mechanisms not linked directly to energy metabolism. Whether this is also the case in working muscle cannot be determined from the present studies although we have preliminary data suggesting that in contracting muscle there is not a shift in NADH levels. However, in that instance there are mechanisms such as release of potassium from the contracting muscle fibers that may be responsible at least in part for the initial phase of vasodilation.

Our conclusions from these studies are based on the assumption that the shift in NADH level is an adequate signal of a change in tissue redox state. It possible that significant shifts in the redox state could occur in the terminal portion of the electron transport chain before such changes take place in the $NAD^+/NADH$ redox couple. However, it seems unlikely that this would occur while myoglobin is fully saturated, which occurs at a value of about 5 mm Hg.

In view of the disparity between the critical PO_2 value found in this study and the normal tissue PO_2, it is possible, as has been suggested by Duling (1978), that oxidases or oxygenases having K_m values closer to the normal range of tissue PO_2 may be involved in blood flow regulation.

12.4 Detection of Increased Oxygen Supply

We turn now to the question of whether there is a specific mechanism by which an overabundance of oxygen delivery is prevented. Certainly, the very potent effect of elevated superfusate oxygen levels on arteriolar tone as well as on blood flow shown in Figure 12.2 puts one in mind of such a direct effect. There have been a number of studies directed toward this question by laboratories other than ours and we will briefly review those findings. It has been proposed that oxygen has a direct effect on vascular smooth muscle itself to cause vasoconstriction.

Evidence in favor of such a mechanism was based on studies in large blood vessels *in vitro* (Detar and Bohr 1968). However, studies of the partial pressure of oxygen in such preparations shows there may be a hypoxic core region in the smooth muscle layer of large vessels *in vitro* that would not normally be present, especially in microcirculatory vessels (Pittman and Duling 1973). Based on these and other studies, it does not appear that the smooth muscle itself is a sensor for PO_2, unless the oxygen levels fall down into the range of critical PO_2 for oxidative metabolism.

Arterioles of hamster cheek pouch are highly sensitive to ambient O_2 levels in *in vivo,* but studies on isolated arterioles from hamster cheek pouch have yielded conflicting results with some vessels showing a constrictor response to oxygen and others being unresponsive (Jackson and Duling 1983). Subsequent studies have shown that lipoxygenase inhibitors abolish the vasoconstrictor effect of oxygen on arterioles of the hamster cheek pouch *in vivo* (Jackson 1988).

The role of prostaglandins in mediating the response to oxygen has been highlighted by studies showing that lowering PO_2 levels increases the release of prostaglandins from endothelial cells (Busse *et al.* 1984). This effect is apparently also present at high PO_2 levels.

It has been shown recently in studies on isolated arterioles from rat cremaster muscle that elevating oxygen above normal tissue levels causes vasoconstriction due to inhibition of release of prostaglandins from the endothelium (Messina *et al.* 1994). The effect is found over a wide range of PO_2 values of 15 to 660 mm Hg, which is in agreement with *in vivo* observations such as shown in Figure 12.2. Such a mechanism is not could provide a means by which elevation of oxygen exerts a vasoconstrictor effect on the arterioles.

In respect to the hypotheses presented in the introduction, these data support the concept of a single mediator of the oxygen effect not linked to energy metabolism, with a reduction in rate of release of a vasodilator substance as oxygen tension increases over a wide range. Since there is a gradient of oxygen tension along the arteriolar network (Duling and Berne 1970), there may also be a gradient within the network of the sensitivity of this mechanism.

12.5 Implications for Blood Substitutes

Finally, these findings have some implications for the area of blood substitutes since it appears that in skeletal muscle one could reduce oxygen delivery by about 50% before reaching the level at which oxidative metabolism becomes compromised. Further studies are required to determine whether in critical organs, such as brain and myocardium, blood flow is also regulated at levels well above that required to maintain oxidative metabolism throughout the organ.

In addition, the information cited above suggests that blood flow and tissue PO_2 are held at normal levels at least in part by a mechanism or mechanisms that operate(s) over a wide range of PO_2. Messina's data suggest that this is sensed at the level of the arteriole, and our data suggest that oxygen tension in resting muscle is maintained well above the critical PO_2 for oxidative energy metabolism where the parenchymal cells might release vasodilator substances.

Blood substitutes that elevate the PO_2 at the level of the arteriole could cause vasoconstriction, a reduction in blood flow and, paradoxically, a decrease in oxygen delivery to the tissues. Since we are dealing with a system of multiple controls, not all of which are known and understood, it would be premature to draw firm conclusions at this time. The available data do suggest, however, that elevating blood PO_2 above normal levels may engage control mechanisms that would tend to offset the advantage of an elevated blood PO_2 in driving oxygen into the tissue by enhancing the gradient for diffusion.

12.6 References

Busse, R., U. Forstermann, H. Masuda, and U. Pohl. The role of prostaglandins in endothelium-mediated vasodilatory response to hypoxia. *Pflugers Arch.* 401: 77-83, 1984.

Chance, B., P. Cohen, F. Jobsis, and B. Schoener. Intracellular oxidation-reduction states *in vivo*. The microfluorometry of pyridine nucleotide gives a continuous measurement of the oxidation state. *Science* 137: 499-508, 1962.

Detar, R., and D.F. Bohr. Oxygen and vascular smooth muscle contraction. *Am. J. Physiol.* 214: 241-244, 1968

Duling, B.R. Oxygen, metabolism and microcirculatory control. In *Microcirculation*, Vol. 2 (G. Kaley and B.M. Altura, eds.). Baltimore: University Park, 1978, pp. 401-429.

Duling, B.R., and R.M. Berne. Longitudinal gradients in periarteriolar oxygen tension. A possible mechanism for the participation of oxygen in the local regulation of blood flow. *Circ. Res.* 27: 669-678, 1970.

Jackson, W.F. Lipoxoygenase inhibitors block O2 responses of hamster cheek pouch arterioles. *Am. J. Physiol.* 255: H711-H716, 1988.

Jackson, W.F., and B.R. Duling. The oxygen sensitivity of hamster cheek pouch arterioles: *In vitro* and *in situ* studies. *Circ. Res.* 53: 515-525, 1983.

Johnson, P.C. Brief Review: Autoregulation of blood flow. *Circ. Res.* 59: 483-495, 1986.

Koller, A., and P.C. Johnson. Methods for mapping and classifying microvascular networks in skeletal muscle. In *Microvascular Networks: Experimental and Theoretical Studies*. Tucson Symposium, Basel: Karger, 1986, pp. 27-37.

Longmuir, I.S. Respiration rate of rat-liver cells at low oxygen concentration. *Biochem. J.* 65: 378-382, 1957.

Messina, E.J., D. Sun, A. Koller, M.S. Wolin, and G. Kaley. Increases in oxygen tension evoke arteriolar constriction by inhibiting endothelial prostaglandin synthesis. *Microvasc. Res.* 8: 151-160, 1994.

Pittman, R.N., and B.R. Duling. Oxygen sensitivity of vascular smooth muscle. I. *In vitro* studies. *Microvasc. Res.* 6: 202-211, 1973.

Prewitt, R.L., and P.C. Johnson. The effect of oxygen on arteriolar red cell velocity and capillary density in rat cremaster muscle. *Microvasc. Res.* 12: 59-70, 1976.

Rosenthal, J., J. La Manna, F.F. Jobsis, J.E. Levasseur, H.S. Kontos, and J.L. Patterson. Effects of respiratory gases on cytochrome a in intact cerebral cortex: is there a critical PO_2? *Brain Res.* 108: 143-154, 1976.

Schubert, R.W., W.J. Whalen, and P. Nair. Myocardial PO_2 distribution: relationship to coronary autoregulation. *Am. J. Physiol.* 234: H361-H370, 1978.

Sparks, H.V. Skin and muscle. In *Peripheral Circulation* (P.C. Johnson, ed.). New York: John Wiley and Sons, 1978. pp. 193-230.

Stainsby, W.N. Autoregulation of blood flow in skeletal muscle during increased metabolic activity. *Am. J. Physiol.* 202: 273-276, 1962.

Torres Filho, I.P., and M. Intaglietta. Microvessel PO_2 measurements by phosphorescence decay method. *Am. J. Physiol.* 265: H1434-H1438, 1993.

Toth, A., M. Tischler, M. Pal, A. Koller, and P.C. Johnson. A multipurpose instrument for quantitative intravital microscopy. *J. Appl. Physiol* 73: 296-306, 1992.

Vanderkooi, J.M., G. Maniara, T.J. Green, and D.F. Wilson. An optical method for measurement of dioxygen concentration based on quenching of phosphorescence. *J. Biol. Chem.* 262: 5476-5482, 1987.

Chapter 13

Tumor Oxygenation and Radiosensitivity

Herman Suit, M.D.

Department of Radiation Medicine, Harvard Medical School,
Massachusetts General Hospital, Boston, Massachusetts 02114

13.1 Introduction

Oxygen is the most potent and thoroughly authenticated modifier of radiation response studied. This pertains to all mammalian cells, tissues and organisms investigated; further, it obtains for all endpoints assessed. The enhancement ratio for oxygen (OER) is 2.5-3.3. This is the ratio of radiation dose administered under hypoxic conditions to that administered under aerobic conditions to produce a defined response. Hypoxia in this context means a PO_2 less than 1 mm Hg and aerobic conditions means a $PO_2 \geq 20$ mm Hg. The radiation sensitizing action of molecular oxygen has been of great interest to radiation oncology and radiation biology as it is maximally sensitizing at physiological concentrations. Further, there is convincing evidence that in some tumors there is a fraction of cells which are hypoxic and viable. This would mean that such cells could be a factor in the failure of radiation to achieve local control in a proportion of such tumors. The central point is that under normal conditions there is a differential PO_2 distribution which favors the tumor relative to the normal tissue with respect to surviving radiation treatments. In reaction to this understanding of tumor biology, there has been extensive and sustained research to devise methods to: measure tissue PO_2, increase tumor PO_2 at irradiation and, hopefully, improve the cure rate of radiation treated patients.

Blood Substitutes: Physiological Basis of Efficacy
Winslow et al., Editors
© Birkhäuser Boston 1995

13.2 Early Observations on Tissue PO_2 and Radiation

Response

In 1921, Holthusen reported that the damaging effect of radiation on ascaris eggs was less in the absence of oxygen (Holthousen 1921). Crabtree and Cramer found (1933) that anaerobiosis protected tumor cells from radiation. Mottram (1935) observed that growth of the beam root (*Vicia faba*) was greater after irradiation in nitrogen than oxygen. This and other findings led to an awareness that PO_2 at the instant of irradiation was an important determinant of the observed response. The oncology community was informed of these developments by a symposium in 1953 organized by L.H. Gray and sponsored by the British Institute of Radiology (Gray *et al.* 1953). Research findings were presented on the magnitude of the "Oxygen Effect" for mammalian cells and animal tissues. O.C.A. Scott demonstrated that the delay of growth of Ehrlich tumor was affected much more by radiation given while the host mice respired O_2 at 3 atmospheres than air; the effect on tumor was much greater than on skin reaction. This provided a basis for optimism that there might be a therapeutic gain from attempts to manipulate tissue PO_2.

13.3 Mechanism of Action of Molecular Oxygen to Increase

Radiation Damage

Molecular oxygen is a highly reactive species that combines with ionized loci in the critical target to produce a local chemical change which the cell has limited ability to repair. Thus, in the presence of oxygen there is a greater likelihood of lethal radiation damage to the cell. To illustrate the marked effect of PO_2 on cell survival, Figure 13.1 is shown. This presents cell survival curves for radiation administered to Chinese hamster CHO cells *in vitro* equilibrated with oxygen concentrations ranging from 0.03% to 100%. There is a 2.7 fold increase in slope of the survival curve as PO_2 is increased from near zero to aerobic levels (Ling *et al.* 1981). The dependence of response on PO_2 occurs predominantly in the range of 1 to about 10 mm Hg; increases in response beyond 20 mm Hg is slight. The mathematical relationship between the OER and pO_2 was developed by Alper and Howard-Flanders (1956) is given by equation 1:

$$(1) \quad OER = m[O_2] + \frac{K}{O_2 + K}$$

m is the maximum OER observed at high PO_2 levels. K is the midpoint of OER between minimum and maximum OER. Experimental data conform closely to the shape predicted by Equation 1. Figure 13.2 presents an excellent and thorough set of experimental data using the CHO cells. In those assays a significant sensitization was observed at O_2 of 0.03% concentration (Ling *et al.* 1981).

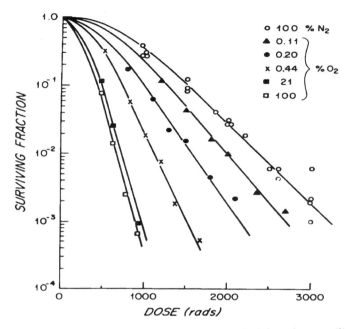

Figure 13.1 Cell survival curves for CHO cells which have been equilibrated with several concentrations of oxygen and then subjected to X radiation (50 kVp) (Ling *et al.* 1981).

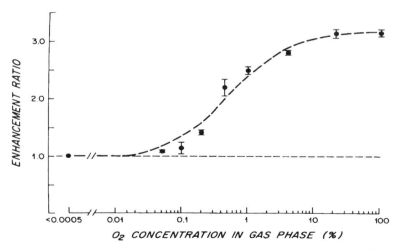

Figure 13.2 OER for increasing concentrations of oxygen for cell killing of CHO cells. (Ling *et al.* 1981).

13.3.1 Measurements of PO_2 in Human Tumors

Polarographic methods have been utilized to measure PO_2 in human tumors since the 1950's (Cater and Silver 1960, Evans and Nylor 1963). These have all shown broad ranges of PO_2 values within individual tumors and, importantly, a greater inter-tumoral heterogeneity in PO_2 levels. The work of Kolstad (1968), Pappova et al. (1982), Gatenby et al. (1988), Höckel et al. (1991) and Vaupel et al. (1991) on carcinomas of the uterine cervix, breast and metastatic cancer to the cervical lymph nodes. Mueller-Kliesser et al. (1981) used cryospectrophotometric techniques to determine the intracapillary oxyhemoglobin saturations in the normal oral mucosa and in adjacent tumor tissue. There was a consistent finding of a substantially lower oxyhemoglobin saturation in the erythrocytes in capillaries of the tumor tissue. Similar findings were obtained for rectal mucosa and cancer by Wendling et al. (1984).

Nitroimadozoles, which act as hypoxic cell sensitizers and have a strong affinity for hypoxic cells, have been shown to bind to cells in human tumors (Urtasun et al. 1986, Franko et al. 1987). This is accepted as good evidence for the presence of hypoxic cells in those tumors. Thus, all of the available data demonstrate that in some, but not all, human solid tumors there is evidence for the presence of hypoxic cells. There is a larger body of data on tumors of experimental animal tumor system which demonstrate the presence of hypoxic cells; these will not be discussed here. An extensive review of this work was published by Moulder and Rockwell (1984). Additional methods have been utilized in some of those studies for indirect estimation of PO_2 values (Olive and Durand 1992, Olive et al. 1993, Boucher et al. 1990, Sevick et al. 1991, Lord et al. 1993, Senger et al. 1993, Shweiki et al. 1992).

13.3.2 Models for Hypoxia in Tumor Tissue

Results of experiments from many laboratories have been interpreted as showing that the OER is: 1) minimally dependent upon cell age, 2) reduced for high LET radiations, and 3) less for radiation given at low dose rate or in small doses/fraction. Further, repair of radiation damage is reduced/absent for hypoxic conditions between irradiations.

There are two categories of hypoxia in tumor tissue: diffusion limited (chronic hypoxia) and perfusion limited (acute hypoxia). The first was described by pathologist H. Thomlinson and biophysicist L.H. Gray in 1955 (Thomlinson and Gray 1955). For their model, the parameters determining PO_2 distribution as a function of distance from a capillary were: PO_2 at the arterial end of the capillary, length of the capillary, blood flow rate through the capillary, oxygen diffusion-coefficient in the tumor, and QO_2 of the intervening tumor cells. The values assigned by Thomlinson and Gray for those parameters for human tissue gave an oxygen diffusion distance of about 160-180 μm. This corresponded well with observed distances from capillaries to necrosis in human lung carcinoma. The diffusion distance of glucose is approximately twice that for oxygen. Thus, the pathological observations were consistent with the concept that the observed necrosis was, at least, in part the consequence of hypoxia.

The second model stemmed from later studies on tumor blood flow which revealed that the flow in tumor capillaries was not constant, but rather there were periods of no flow in a fraction of the capillaries (Brown 1979). The "no flow" fraction might be fairly constant, but the actual capillaries involved would be continuously changing. This meant that at any time there would be some regions for which PO_2 was perfusion-limited. Thus, the PO_2 of cells adjacent to and downstream from the start of the no-flow length of capillary would rapidly go to near zero. As the "no flow" was temporary, necrosis would not necessarily develop. According to the second model, there would be episodic hypoxia over largish areas; these would likely far exceed the oxygen diffusion length. The dynamics of development of hypoxia was quite different for the two models.

13.4 Strategies Employed to Minimize the Importance of

Hypoxic Cells in Radiation Therapy

There has been an impressive array of strategies conceived and employed in the clinic to reduce the importance of the hypoxic tumor cells on the outcome of radiation treatments. These include:

- Respiration of oxygen at 1 or 3 atmospheres or carbogen at 1 atmospheres.

- Radiation under conditions of local hypoxia (application of tourniquet on the proximal extremity in treatment of a distal extremity sarcoma); the rationale being to eliminate the differential in tissue PO_2, which normally protects some of the cells.

- High LET radiation (fast neutrons, neon ion, boron neutron neutron capture therapy) as the OER for these particle beams is lower than for photons.

- Hypoxic cell sensitizers (metronidazole, misonidazole, nimorazole, etanidazole, RSU-1069, SR4322, *etc.*).

- Hyperthermia.

- Agents to increase tumor blood flow, *e.g.*, nicotinamide, calcium channel blockers.

- Transfusion; hemoglobin substitutes (perfluorochemical compounds, *e.g.*, Fluorosol); erythropoeitin.

- Agents that modify the hemoglobin O_2 affinity.

- Agents specifically toxic to hypoxic cells.

- Agents that suppress O_2 utilization.

- Intra-arterial infusion of H_2O_2.

There has been some degree of success in laboratory animal tumor systems with all of these approaches. The clinical applications have met with only modest success.

13.5 Experimental Animal Tumor Studies

There is an extensive literature on this subject. The study of Powers and Tolmach (1963) showed that there was a population of cells within a mouse lymphoma which were hypoxic in their response to radiation (see Figure 13.3). The cell survival curve exhibited two components. The ratio

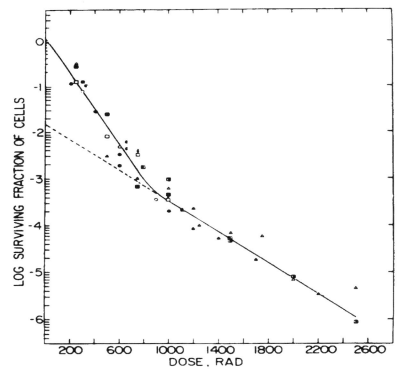

Figure 13.3 Cell survival curve for single dose irradiation of a mouse lymphoma demonstrating a two component curve (Powers and Tolmach 1963).

of slopes of the two components corresponded to the OER for mammalian cells, with the D_0 of the resistant tail being 2.6 Gy, *viz.*, the value expected for hypoxic cells.

Brief reference will be made here to several studies on early generation isotransplants of the mammary carcinoma of the C3H/Sed mouse, MCaIV. The dose response curve for local control of tumor can be sharpy modified by alteration in the oxygenation of the tumor at the time of irradiation. This is shown by the curves in Figure 13.4, which were derived from assays performed on host mice respiring air under normal conditions, respiration of O_2 at 4 atmospheres or clamp hypoxia [Moulder and Rockwell 1984]. An indication of the proportion of cells which are hypoxic can be derived from the ratio (TCD_{50} Hypoxic/TCD_{50} Air) which increases with tumor size. (TCD_{50} is the radiation dose which achieves eradication of the tumor in 50% of the irradiated lesions.)

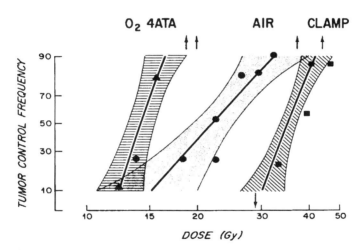

RADIATION DOSE TUMOR CONTROL RESPONSE ASSAYS FOR 0.6 mm³ MCa IV (v = 1, 250 kvp xrays)

Figure 13.4 Dose response curves for single dose irradiation of small isotransplants of a C3H mouse mammary carcinoma under one of three conditions: 1) normal Air breathing and blood flow, 2) respiring O_2 at 4ATA, or 3) local tissue hypoxia. (Shaded area: 95% confidence bands) (Suit and Maeda 1967)

Table 13.1 Tumor hypoxia and size of MCaIV.

TCD$_{50}$	Hypoxia/Air
(single dose)	
Microcolonies	1.8
0.6 mm^3	1.6
250 mm^3	1.1

(Suit and Maeda 1967)

For many experimental tumor systems, the enhancement ratio for O_2 at 3 atmospheres is greater for fractionated than single-dose irradiation. This was found for MCaIV, as indicated by the results in Table 13.2 (Howes and Suit 1974).

Table 13.2 Enhancement ratios (ER) for O_2 3 atmospheres and fractionation of the radiation treatment.

TCD$_{50}$	AIR/O$_2$ 3 Atm
n =1	1.13
n =10	1.56 (27°C)
	2.18 (35°C)
n =15	1.72 (35°C)

(Suit, Howes and Hunter 1977)

The ER (TCD$_{50}$ Air/Test) for respiration for different times of O_2 at 1 or 3 Atm or carbogen at 1 Atm has been examined in this system for radiation given in 10 equal doses (n = 10). The findings are that respiration for 5 minutes yields a near maximum effect, see Table 13.3 (Suit *et al.* 1972).

Table 13.3 ERs for respiration of O_2 at 1 or 3 Atm or carbogen for various times.

Respired Gas	Times	ER
O$_2$ 1 Atm	0.05 min	1.09
	15 min	1.41
	60 min	1.22
Carbogen	0.5 min	0.99
	15 min	1.34
O$_2$ 3 Atm	0.05 min	1.76
	15 min	1.79

(Suit, Marshall and Woerner 1972)

An experiment was performed to derive the Therapeutic Gain Factor (TGF) for respiration of O_2 3 Atm. (The TGF is the ER for tumor/normal tissue; in this assay, normal tissue was the intestinal tract.) Radiation was administered in 5 or 15 equal-dose fractions over a total time of 5 days. There was a clear gain for both fractionation schedules, *viz.*, TGFs were 1.5-1.7, (Powers and Tolmach 1963); in the same experiments a postive TGF was not obtained for fast neutron irradiations. There is a major interest in the combination of carbogen and nicotinamide to improve the radiation response based upon this rationale: carbogen to decrease the hypoxic regions by diffusion of oxygen, and the nicotinamide to improve the perfusion. For some tumor system, there is a greater ER for carbogen than for oxygen. The results of studies by Rojas *et al.* (1990) at the Gray Lab have been quite positive for the CaNT tumor.

Table 13.4 ERs for oxygen, carbogen alone or combined with nicotinamide in treatment of tumor CaNT. Radiation was given as 10 fractions over 5 days. ER was found to be maximal at 5 minute breathing times from the tumor.

Oxygen Condition	ER
Air	1.16
O_2 1 Atm	1.52
Carbogen	1.82
Carbogen + Nicotinamide	
0 mg	1.5
100 mg	1.6
200 mg	1.7
300 mg	1.7
500 mg	1.8

13.6 Correlation Between the Response of Human Tumors and Measured PO_2 Values

There have been three reports which give follow-up status after radiation treatment and the pre-treatment PO_2 measured values. These were by Kolstad (1968), Gatenby *et al.* (1988) and Höckel *et al.* (1991). These are encouraging in that in each of the three studies, the tumors which had large numbers of measurements indicating hypoxic regions were much more likely to regrow. For example, in Gatenby's study of cervical lymph nodes involved by metastatic carcinoma, the cut-off was the proportion of measured PO_2 values that were <8 mm Hg. Complete responses were observed in 1/11 and 19/19 tumors for which ≥26% or <26%

of measured PO_2 values were <8 mm Hg, respectively. Höckel *et al.* (1991) found a much higher survival rate in patients whose cervical carcinoma exhibited only slight hypoxia; the difference was significant. Those reported results are being maintained and also for the subset of patients who were treated by radiation alone (M. Höckel, personal communication, 1994).

13.6.1 Clinical Trials of Modifiers of Tissue PO_2

Overgaard and Horsman (1993) have performed a meta-analysis on the large number of Phase III trials of respiration of oxygen at increased pressure, administration of one of the hypoxic cell sensitizers. The result was a small, but significant, benefit from the modifier.

Henk, Kunkler and Smith (1977) and Henk and Smith (1977) reported results from two successive trials of O_2 3 Atm for carcinomas of the head/neck region. Both were highly significant in favor of the O_2 3 Atm arms for local control. In one, a significant gain was also seen for survival. Bush *et al.* (1978) obtained a gain from transfusion of patients with carcinoma of the uterine cervix who were anemic.

13.6.2 Clinical Problem

The clinician needs means for assessing the physiological status of the tumor in the individual patient with special reference to PO_2. The limited results mentioned are indeed promising but not adequate for regular use of a PO_2 modifier. A serious concern is that the clinical trials to date have been marred by the accession into the trial all patients who had tumors of a defined site, stage and pathological type. However, all of the tumor PO_2 values measured polarographically indicate that there is an important degree of inter-tumoral heterogeneity in PO_2 distributions. Accordingly, the optimal trial design would have a demonstrated low PO_2 as an eligibility criterion. Were this to obtain, there is a sound basis for anticipating a much greater gain from the employment of a procedure which modifies tumor PO_2. Thus, additional clinical data are required as to the degree of correlation between PO_2 and outcome of radiation treatment and then to procede with trials of one of the modifiers.

However, there is an obvious need for a technical means for determining tissue PO_2 that is simpler, more readily tolerated by patients and less costly and cumbersome. The interest in tumor PO_2 extends beyond the effort to enhance efficacy of radiation treatments. The physiological features that impede access of O_2 to the tumor cells also decrease the access of chemotherapeutic agents, the diverse components of the immune rejection reaction, vectors for gene therapy, *etc.* (Jain 1990). This is an important clinical problem and one with real challenges to the clinically oriented and the basic physiologist.

13.7 References

Alper, T., and P. Howard-Flanders. The role of oxygen in modifying the radiosensitivity of *E. coli* B. *Nature* 178: 978-979, 1956.

Boucher, Y., L.T. Baxter, and R.K. Jain. Interstitial pressure gradients in tissue-isolated and subcutaneous tumors: implications for therapy. *Cancer Res.* 50: 4478-4484, 1990.

Brown, J.M. Evidence for acutely hypoxic cells in mouse tumors and possible mechanism. *Brit. J. Radiol.* 52: 650-656, 1979.

Bush, R.S., R.D.T. Jenkin, W.E.C. Allt, *et al.* Definitive evidence of hypoxic cells influencing cure in cancer therapy. Br. *J. Cancer* 37: 302-306, 1978.

Cater, D.B., and I.A. Silver. Quantitative measurements of oxygen tension in normal tissues and in the tumors of patients before and after radiotherapy. *Acta Radiol.* 53: 233-256, 1960.

Crabtree, H.G., and W. Cramer. The action of radium on cancer cells. II. Some factors determining the susceptibility of cancer cells to radium. *Proc. Roy. Soc.* B. 113: 226-238, 1933.

Evans, N.T.S., and P.F.D. Naylor. The effect of oxygen breathing and radiotherapy upon tissue oxygen tension of some human tumors. *Br. J. Radiol.* 36: 418-425, 1963.

Franko, A.J., C.J. Koch, B.M. Garrecht, J. Sharplin, and D. Hughes. Oxygen dependance of binding of misonidazole to rodent and human tumors *in vitro. Cancer Res.* 47: 5367-5376, 1987.

Gatenby, R.A., H.B. Kessler, J.S. Rosenblum, L.R. Coia, P.J. Moldofsky, W.H. Hartz, and G.J. Broder. Oxygen distribution in squamous cell carcinoma metastases and its relationship to outcome of radiation therapy. *Int. J. Radiat. Oncol. Biol. Phys.* 14: 831-838, 1988.

Gray, L.H., A.S.D. Conger, M. Ebert, S. Hornsey, and O.C.A. Scott. The concentration of oxygen dissolved in tissue at the time of irradiation as a factor in radiotherapy. *Br. J. Radiol.* 26: 638-648, 1953.

Henk, J.M., P.B. Kunkler, and C.W. Smith, C.W. Radiotherapy and hyperbaric oxygen in head and neck cancer. Final report of first controlled clinical trial. *Lancet* 7: 101-103, 1977.

Henk, J.M., and C.W. Smith. Radiotherapy and hyperbaric oxygen in head and neck cancer. Final report of first controlled clinical trial. *Lancet* 7: 104-105, 1977.

Höckel, M., K. Schlenger, C. Knoop, and P. Vaupel. Oxygenation of carcinomas of the uterine cervix: Evaluation by computerized O_2 tension measurements. *Cancer Res.* 51: 6098-6102, 1991.

Holthousen, H. Beitrage zur biologie der strahlenwirkung. Untersuchugen an askarideneiern. *Pflüg. Arch. ges. Physiol.* 187: 1, 1921.

Howes, A.E., and H.D. Suit. The effect of time between fractions on the response of tumors to irradiation. *Radiat. Res.* 57: 342-348, 1974.

Jain, R.K. Physiological barriers in the delivery of monoclonal antibodies and other macromolecules in tumors. *Cancer Res.* 50: 814-819, 1990.

Kolstad, P. Intercapillary distance, oxygen tension, and local recurrence in cervix cancer. *Scand. J. Clin. Lab. Invest.* Suppl. #106: 145-157, 1968.

Ling, C.C., H.B. Michaels, L.E. Gerweck, E.R. Epp, and E.C. Peterson. Oxygen sensitization of mammalian cells under different irradiation conditions. *Radiat. Res.* 86: 325-340, 1981.

Lord, E.M., L. Harwell, and C.J. Koch. Detection of hypoxic cells by monoclonal antibody recognizing 2-nitromidazole adducts. *Cancer Res.* 53: 5721-5726, 1993.

Mottram, J.C. On the alteration in the sensitivity of cells towards radiation produced by cold and by anaerobiosis. *Br. J. Radiol.* 8: 32, 1935.

Moulder, J.E., and S. Rockwell. Hypoxic fractions in solid tumors: experimental techniques, methods of analysis, and survey of existing data. *Int. J. Radiat. Oncol. Biol. Phys.* 10: 695-712, 1984.

Mueller-Klieser, W., P. Vaupel, R. Manz, and R. Schmidseder. Intracapillary oxyhemoglobin saturation in malignant tumors in humans. *Int. J. Radiat. Oncol. Biol. Phys.* 7: 1397-1404, 1981.

Olive, P.L., and R.E. Durand. Detection of hypoxic cells in a murine tumor with the use of comet assay. *J. Natl. Cancer Inst.* 84: 707-711, 1992.

Olive, P.L., R.E. Durand, J. LeRiche, and S.M. Jackson. Gel electrophoresis of individual cells to quantify hypoxic fractions in human breast cancers. *Cancer Res.* 53: 733-736, 1993.

Overgaard, J., and M.R. Horsman. Overcoming hypoxic cell radioresistance. In *Basic Clinical Radiobiology for Radiation Oncologists* (G.G. Steel, ed.) London: Edward Arnold, 1993, pp. 163-172.

Poppova, N., E. Siracka, A. Vacek, and J. Durkovsky. Oxygen tension and predication of therapy response. Polarographic study in human breast cancer. *Neoplasma* 29: 669-674, 1982.

Powers, W.E., Tolmach, L.J. Survival of 6C3HED mouse lymphosarcoma cells irradiated in vivo. Initial D_0 1.1 Gy, final D_0 2.6 Gy. *Nature* 197: 710-711, 1963.

Rojas, A., U. Carl, and K. Reghebi. Effect of normobaric oxygen on tumor radiosensitivity. Fractionated studies. *Int. J. Radiat. Oncol. Biol. Phys.* 18: 547-553, 1990.

Shweiki, D., A. Itin, D. Soffer, and E. Keshet. Vascular endothelial cell growth factor induced by hypoxia may mediate hypoxia-initiated angiogenesis. *Nature* 359: 843-845, 1992.

Senger, D.R., L. Van de Water, L.F. Brown, *et al*. Vascular permeability factor (VGF, VEGF) in tumor biology. *Cancer Metast. Rev.* 12: 303-324, 1993.

Sevick, E.M., B. Chance, J. Leigh, *et al*. Quantification of time and frequency resolved optical spectra for the determination of tissue oxygenation. *Anal. Biochem.* 195: 330-351, 1991.

Suit, H.D., and M. Maeda. Hyperbaric oxygen and radiobiology of a C3H mouse mammary carcinoma. *J. Natl. Cancer Inst.* 39: 639-652, 1967.

Suit, H.D., N. Marshall, and D. Woerner. Oxygen, oxygen plus carbon dioxide and radiation therapy of a mouse mammary carcinoma. *Cancer* 30: 1154-1158, 1972.

Thomlinson, R.H., and L.H. Gray. The histological structure of some human lung cancers and the possible implications for radiotherapy. *Br. J. Cancer* 9: 539-549, 1955.

Urtason, R.C., J.D. Chapman, J.A. Raleigh, A.J. Franko, and C.J. Koch. Binding of ^{3}H-misonidazole to solid human tumors as a measure of tumor hypoxia. *Int. J. Radiat. Oncol. Biol. Phys.* 12: 1263-1267, 1986.

Vaupel, P., K. Schlenger, C. Knoop, and M. Hckel. Oxygenation of human tumors: evaluation of tissue oxygen distribution in breast cancers by computerized O_2 tension measurements. *Cancer Res.* 51: 3316-3322, 1991.

Wendling, P., R. Manz, G. Thews, and P. Vaupel. Heterogeneous oxygenation of rectal carcinomas in humans: a critical parameter for preoperative irradiation? *Adv. Exp. Med. Biol.* 180: 293-300, 1984.